Practical Guide to
Large Database
Migration

Preston Zhang

CRC Press
Taylor & Francis Group
Boca Raton   London   New York

CRC Press is an imprint of the
Taylor & Francis Group, **an informa** business

A SCIENCE PUBLISHERS BOOK

CRC Press
Taylor & Francis Group
6000 Broken Sound Parkway NW, Suite 300
Boca Raton, FL 33487-2742

Printed on acid-free paper
Version Date: 20181226

International Standard Book Number-13: 978-1-1383-9162-8 (Hardback)

Library of Congress Cataloging-in-Publication Data

Names: Zhang, Preston, author.
Title: Practical guide to large database migration / Preston Zhang,
  University of Georgia, Database Administrator Watkinsville, Georgia, USA.
Description: Boca Raton, FL : CRC Press, Taylor & Francis Group, [2019] |
  Includes bibliographical references and index.
Identifiers: LCCN 2018060369 | ISBN 9781138391628 (hardback : acid-free paper)
Subjects: LCSH: Systems migration. | Database management.
Classification: LCC QA76.9.S9 Z53 2019 | DDC 005.74--dc23
LC record available at https://lccn.loc.gov/2018060369

Visit the Taylor & Francis Web site at
http://www.taylorandfrancis.com

and the CRC Press Web site at
http://www.crcpress.com

Preface

The success of any modern business relies on its ability to adapt to fast changing business environments: for example, IT infrastructure, database version and servers need to be upgraded every few years.

When upgrading existing hardware or servers or to a new system, a common task for an IT team is how to migrate database from old system to new system. We need to make sure that there is low business downtime during the migration. We also need to test new system to avoid application failure.

As a database administrator I have done database migrations many times. For example, migrating a MySQL database from a developer server to a production server or migrating a SQL Server database from an older server to a new server. Usually, migrating data within the same database system is not a difficult task. However, it is a challenging when migrating database from one system to a different system. For example, migrating a SQL Server database to Oracle database system.

I have tried to find a book about database migration in Oracle, SQL Server and MySQL database systems, but I only found books about Oracle or AWS database migration. I would like to share database migration experience with database developers and DBAs. All the migration examples in this book use large databases in Oracle, SQL Server and MySQL. This book is written in easy to read style with step-by-step examples.

Who This Book Is For

This book is for intermediate database developers, database administrators. If you are not familiar with SQL syntax and installation of Oracle/SQL Server/MySQL please read my book "Practical Guide to Oracle SQL, T-SQL and MySQL". The SQL code in this book is fully tested in Oracle 11g, Oracle 12c, SQL Server 2012, SQL Server 2016, MySQL 5.5 and MySQL 5.7.

How to Use this Book

To run the examples from this book you need to install the following database systems and development tools:

> Oracle 11g, Oracle 12c
> Oracle SQL Developer
> SQL Server 2012, SQL Server 2016
> SQL Server Management Studio 2012 or above
> MySQL Server 5.5, MySQL Server 5.7
> MySQL Workbench 6.3

All the above software can be download from Oracle.com, Microsoft.com and MySQL.com

Contents

Chapter 1
Introduction to Database Migration

1.1 What is Database Migration

Database migration is the process of moving a database from a vendor to another or upgrading current version of database software. There are many reasons why organizations would want to migrate their databases:

- High cost of traditional database platform ownership and maintenance cost
- Replacing legacy servers
- Updating storage equipment
- Company mergers
- Moving data to a cloud provider
- Application migration

According to DB-Engines, in June 2018, the most widely used systems are Oracle, MySQL, Microsoft SQL Server, PostgreSQL, IBM DB2, Microsoft Access, and SQLite. All the migration examples in this book use SQL Server, MySQL, Oracle and Access because those are the top databases in the world.

Database migration is not an easy task. For example, it's quite a challenge to migrate Oracle database. This book provides step by step guides for migrating databases between SQL Server, MySQL, Oracle and Microsoft Access. I hope that this book can help database developers and DBAs to make database migration easier.

1.2 Database Migration Stages

Database migration usually involves several stages:

- Migration Preparing Stage
- Data Migration Stage
- ETL Stage
- Database and Application Testing Stage

1.2.1 Migration Preparing Stage

- Identify the source database tables, views, stored procedures and data
- Identify application and database users on the source system

- Identify privileges and permissions on the source system
- Find application software that uses the source database
- Estimate migration risks and cost
- Check hardware capacity
- Evaluate migration tools

1.2.2 Data Migration Stage

Schema mapping between source database and target database. Data type mapping is easy if there is no change in database format. For example, database migration from MySQL to MySQL. However, data type mapping is a difficult task for large tables when migrating from one database format to another database format. For example, database migration from SQL Server to Oracle.

Database migration tools can help originations to accomplish migration projects. The tools usually migrate views to tables. It's hard to find a tool that can migrate triggers and stored procedures. DBAs or database developers can create views in target database by using the migrated tables. They also can generate SQL code for views from source database then apply the code to target database. For triggers and stored procedures DBAs and database developers need to get SQL code from the source database then convert the SQL code for the target database.

Below is a data types mapping table. You can use this table to convert data types manually or verify data types after database migration. In later chapter you will see that database migration tools can map date types between a source and a target database automatically.

Table 1-1 Data types mapping table

SQL Server Type	MySQL Type	Oracle Type	Access Type
INT	INT	NUMBER(10)	LONG INTEGER
TINYINT	TINYINT	NUMBER(3)	BYTE
SMALLINT	SMALLINT	NUMBER(5)	INTEGER
BIGINT	BIGINT	NUMBER(20)	
BIT	TINYINT(1)	NUMBER(3)	YES/NO
FLOAT	FLOAT	FLOAT(53)	DECIMAL
REAL	FLOAT	FLOAT(24)	SINGLE
NUMERIC	DECIMAL	NUMBER(p, s)	CURRENCY
DECIMAL	DECIMAL	NUMBER(p, s)	CURRENCY
MONEY	DECIMAL	NUMBER(19, 4)	CURRENCY
SMALLMONEY	DECIMAL	NUMBER(10, 4)	
CHAR	CHAR	CHAR	
NCHAR	CHAR/LONGTEXT	NCHAR	
VARCHAR	VARCHAR	VARCHAR2	MEMO
NVARCHAR	VARCHAR	NVARCHAR	TEXT

Table 1-1 contd. ...

... Table 1-1 contd.

SQL Server Type	MySQL Type	Oracle Type	Access Type
DATE	DATE		
DATETIME	DATETIME	DATE	
SMALLDATETIME	DATETIME	DATE	DATE/TIME
TIME	TIME		
TIMESTAMP	TIMESTAMP	TIMESTAMP	
BINARY	BINARY	RAW	BINARY(SIZE)
VARBINARY	VARBINARY	RAW	
TEXT	VARCHAR	LONG	MEMO
NTEXT	VARCHAR	LONG	
IMAGE	LONGBLOB	LONG RAW	
XML	TEXT	LONG	

1.2.3 ETL (Extract, Transform, and Load) Stage (Optional)

- Sometimes organizations want to extract data from several sources and transform the data based on business rules then load the data into target databases.
- Transforming process may include sorting, joining data and cleaning data, etc.
- Loading process includes loading the transformed data into target databases or data warehouses.

1.2.4 Database and Application Testing Stage

- Verify the migrated data on target database
- Create the same application and database users on the target system
- Create the same privileges and permissions on the target system
- Test views, triggers, stored procedures and functions on the target database
- Test the database and application by the end users to make sure there are no errors

1.3 Database Migration Paths

You will see database migration examples from SQL Server to MySQL, SQL Server to SQL Server and SQL Server to Oracle in Chapter 2. We will use SQL Server **Chinook** database in this book.

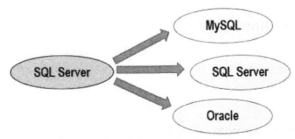

Figure 1-1 SQL Server migration paths

You will see database migration examples from MySQL to MySQL, MySQL to SQL Server and MySQL to Oracle in Chapter 4. We will use MySQL **Employees** database in this book.

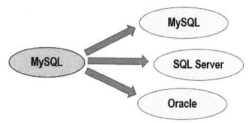

Figure 1-2 MySQL database migration paths

You will see database migration examples from Oracle to MySQL, Oracle to SQL Server and Oracle to Oracle in Chapter 6. We will use Oracle **HR** or **SH** database in this book.

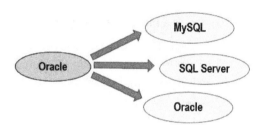

Figure 1-3 Oracle database migration paths

You will see database migration examples from Microsoft Access to MySQL, Microsoft Access to SQL Server and Microsoft Access to Oracle in Chapter 5. We will use Microsoft Access **MonthlySalesReports** database in this book.

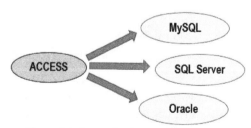

Figure 1-4 Access database migration paths

1.4 Database Migration Tools

There are a lot of database migration tools available. I have tested the following tools and I will use those tools to demonstrate large database migration in this book.

- ESF Database Migration Toolkit
- Data Loader
- MySQL Workbench Migration Wizard

- SQL Server Migration Assistant for MySQL
- SQL Server Migration Assistant for Oracle
- SQL Server Migration Assistant for Access

1.4.1 ESF Database Migration Toolkit

Website: https://www.easyfrom.net

ESF Database Migration Toolkit can help companies migrate data between Oracle, MySQL, MariaDB, SQL Server, PostgreSQL, IBM DB2, SQLite, Microsoft Access, etc. It can migrate most database objects except procedure, function and trigger. Database view will be migrated to table. The trial version has a migration limit for 50,000 rows per table and it will add an extra field in tables. That is good enough for testing database migration.

ESF Database Migration Toolkit is very easy to use. It saves database administrators a lot of time for database migration projects.

1.4.2 Data Loader

Website: https://dbload.com

- Data Loader works well for database migration, but the trial version has 50 rows limit per table, that's why I will not use it as a migration tool in this book.
- Data Loader Standard Edition has 100,000 rows limit per table.
- Data Loader Professional Edition has 1,000,000 rows limit per table.
- Data Loader Enterprise Edition has unlimited rows per.

Data Loader supports MySQL, Oracle, MS SQL Server, MS Access, Excel, FoxPro/DBF, CSV/text files:

Product	Single License	2-5 Licenses	6 & above Licenses	FastSpring Store	Share-it! Store
Standard Edition	$99	$79 per license	$69 per license	BUY NOW	BUY NOW
Professional Edition	$199	$179 per license	$159 per license	BUY NOW	BUY NOW
Enterprise Edition	$299	$279 per license	$259 per license	BUY NOW	BUY NOW

Figure 1-5 Data Loader different versions

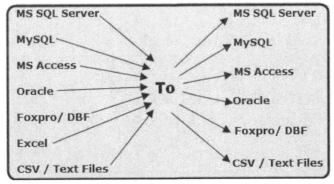

Figure 1-6 Data Loader database migration paths

1.4.3 *MySQL Workbench Database Migration Wizard*

The MySQL Workbench Migration Wizard allow users to convert an existing database to MySQL in few steps. The source database might be MySQL, SQL Server, PostgreSQL and Microsoft Access, etc.

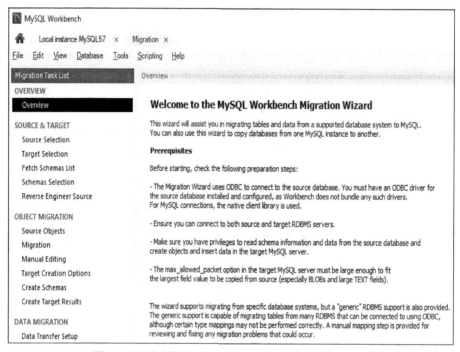

Figure 1-7 MySQL Workbench Database Migration Wizard

1.4.4 *SQL Server Migration Assistant Tools*

Microsoft SQL Server Migration Assistant (SSMA) is a tool to help database migration to SQL Server from MySQL, Oracle, Microsoft Access, DB2 and Sybase DB. We will use SQL Server Migration Assistant for MySQL and SQL Server Migration Assistant for Oracle in this book.

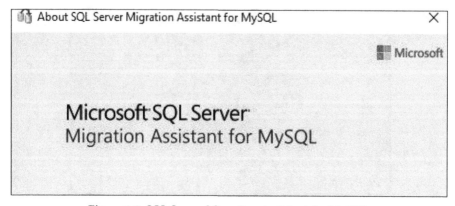

Figure 1-8 SQL Server Migration Assistant for MySQL

Figure 1-9 SQL Server Migration Assistant for Oracle

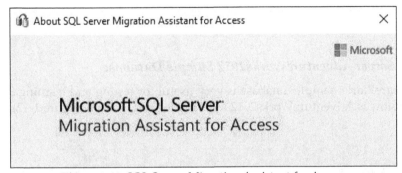

Figure 1-10 SQL Server Migration Assistant for Access

1.5 Sample Large Databases

1.5.1 SQL Server Chinook Sample Database

The Chinook Database is a sample database for SQL Server, Oracle, MySQL etc. It is being used for application prototypes and website backend database. It includes the following tables: Artists, Albums, Employee, Track, Playlist, Playlisttrack, Invoice, Invoiceline, Genre, Customer and Mediatype. The maximum rows is 8,715 rows (Playlisttrack table).

Download the Chinook Database below:
https://github.com/jimfrenette/chinook-database

Table Name	# Records
dbo.Playlisttrack	8,715
dbo.Track	3,503
dbo.Invoiceline	2,240
dbo.Invoice	412
dbo.Album	347
dbo.Artist	275
dbo.Customer	59
dbo.Genre	25
dbo.Playlist	18
dbo.Employee	8
dbo.Mediatype	5

Figure 1-11 Tables in Chinook database

1.5.2 SQL Server AdventureWorks2012 Sample Database

The AdventureWorks sample database is very useful for testing and learning SQL Server database. Below is AdventureWorks2012 table names and records (Figure 1-12):

Table Name	# Records
Sales.SalesOrderDetail	121,317
Production.TransactionHistory	113,443
Production.TransactionHistoryArchive	89,253
Production.WorkOrder	72,591
Production.WorkOrderRouting	67,131
Sales.SalesOrderHeader	31,465
Sales.SalesOrderHeaderSalesReason	27,647
Person.BusinessEntity	20,777
Person.EmailAddress	19,972
Person.Password	19,972
Person.Person	19,972
Person.PersonPhone	19,972
Sales.Customer	19,820
Person.Address	19,614
Person.BusinessEntityAddress	19,614
Sales.CreditCard	19,118

Figure 1-12 Tables in AdventureWorks2012 database

1.5.3 *MySQL Employees Sample Database*

The Employees sample database provides a large database with size about 160MB. It has 4 million records in total.

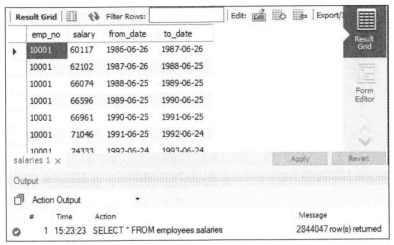

Figure 1-13 Sample records from MySQL Employees database salaries table

1.5.4 *MySQL Sakila Sample Database*

The Sakila sample database includes MySQL new features. It is good for tutorials and examples. Figure 1-17 shows views, stored procedures and functions in Sakila database.

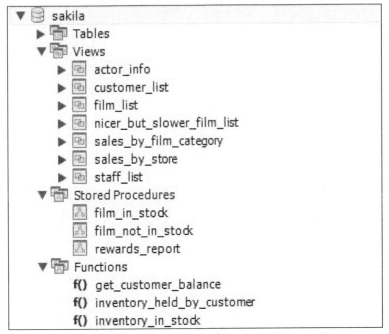

Figure 1-14 Sakila database structures

1.5.5 Oracle Human Resources (HR) Sample Database

Human Resources (HR) database is useful for introducing fundamental database topics. Figure 1-15 shows that dept_emp table has 331,613 rows.

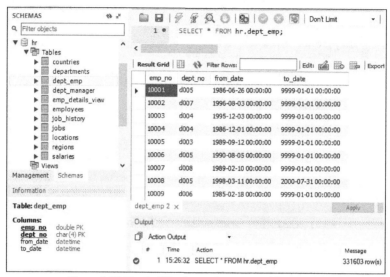

Figure 1-15   Oracle HR schema dept_emp table records

1.5.6 Oracle Sales History (SH) Sample Database

The sample database has many reports that evaluate past data trends. There are annual, quarterly, monthly, and weekly sales reports. The database also has data for sales by geographical area.

	TABLE_NAME	COUNT
1	SALES	918843
2	COSTS	82112
3	CUSTOMERS	55500
4	FWEEK_PSCAT_SALES_MV	11266
5	SUPPLEMENTARY_DEMOGRAPHICS	4500
6	TIMES	1826
7	PROMOTIONS	503
8	PRODUCTS	72
9	CAL_MONTH_SALES_MV	48
10	COUNTRIES	23
11	DRSUP_TEXT_IDXR	22
12	CHANNELS	5

Figure 1-16   Oracle Sh schema tables

1.5.7 Microsoft Access MonthlySalesReports Sample Database

Figure 1-17 shows that tblOrderDetails has 121,317 records.

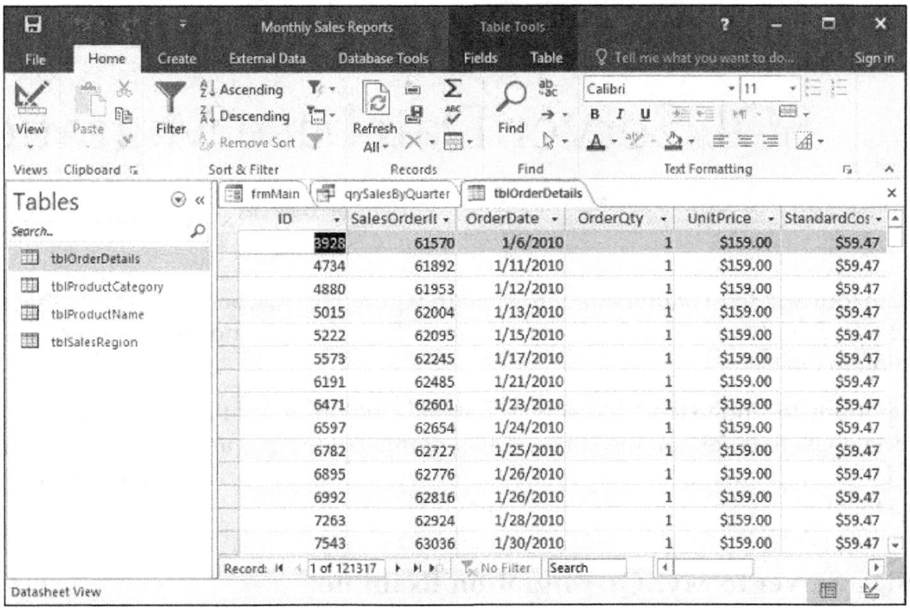

Figure 1-17 Tables in Access MonthlySalesReports database

Summary

Chapter 1 covers the following:

Database migration definition

Database migration stages including Migration Preparing Stage, Database Migration Stage, ETL Stage and Database and Application Testing Stage

Database migration sample paths

Database migration tools for Oracle SQL Server, MySQL and Microsoft Access databases

Database migration sample databases in Oracle, SQL Server, MySQL and Microsoft Access

Chapter 2
SQL Server Database Migration

The main topics in this chapter are illustrated in Figure 2-1: SQL Server database migration to MySQL, SQL Server database migration to SQL Server and SQL Server database migration to Oracle.

It's important to understand the source database before migration. You can open SQL Server database and view all the tables, views, stored procedures and functions. A decision needs to be made weather your organization want to keep current database objects or change them.

2.1 SQL Server to MySQL Migration Example

We will use ESF Database Migration Toolkit and MySQL Workbench Migration Wizard in this section.

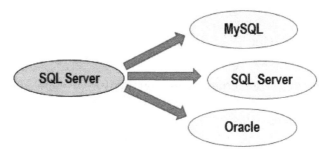

Figure 2-1 SQL Server database migration paths

2.1.1 Using ESF Database Migration Toolkit

Prerequisites:	Windows XP/Vista/7/8/8.1/10
	SQL Server 6.5 and above
	MySQL 3.23 and above
Source Server and Database:	SQL Server 2016 Chinook database
Target Server and Database:	MySQL 5.7 Chinook database
Migration Tool:	ESF Database Migration Toolkit
Database Transform:	Yes

Steps to migrate SQL Server to MySQL Using ESF Database Migration Toolkit:

Step 1:

Let us set up Chinook database on the source SQL Server. If you have not downloaded the Chinook.sql file, please see Chapter 1 for the download link. Open the Chinook.sql file in Query window and execute the commands. The Chinook database appears in the left pane after refreshing the Object Explorer:

```
/****************************************************************************
    Chinook Database - Version 1.4
    Script: Chinook_SqlServer.sql
    Description: Creates and populates the Chinook database.
    DB Server: SqlServer
    Author: Luis Rocha
    License: http://www.codeplex.com/ChinookDatabase/license
 ****************************************************************************/


/****************************************************************************
    Drop database if it exists
 ****************************************************************************/
IF EXISTS (SELECT name FROM master.dbo.sysdatabases WHERE name = N'Chinook')
BEGIN
        ALTER DATABASE [Chinook] SET OFFLINE WITH ROLLBACK IMMEDIATE;
        ALTER DATABASE [Chinook] SET ONLINE;
        DROP DATABASE [Chinook];
END

GO

/****************************************************************************
    Create database
 ****************************************************************************/
CREATE DATABASE [Chinook];
GO

USE [Chinook];
GO
```

Figure 2-2 Opening Chinook.sql in a query window

Step 2:

Open table columns you will notice that all the fields have double quotes and brackets. Remove the double quotes and brackets from the filed names.

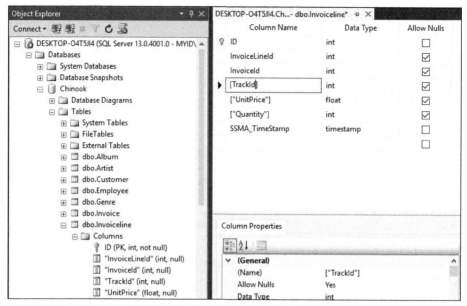

Figure 2-3 Removing the double quotes and brackets from the filed names

Step 3:

Right Click **Chinook** database and select **Reports -> Standard Reports -> Disk Usage by Table:**

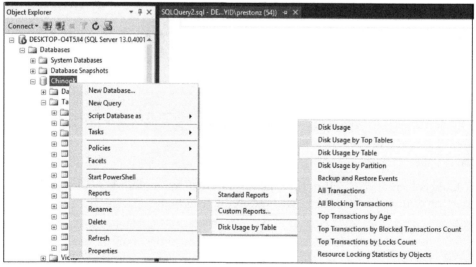

Figure 2-4 Disk Usage by Table Report

Step 4:

Figure 2-5 shows all the table records from **Disk Usage by Table** Report:

Table Name	# Records
dbo.PlaylistTrack	8,715
dbo.Track	3,503
dbo.InvoiceLine	2,240
dbo.Invoice	412
dbo.Album	347
dbo.Artist	275
dbo.Customer	59
dbo.Genre	25
dbo.Playlist	18
dbo.Employee	8
dbo.MediaType	5

Figure 2-5 Disk Usage by Table Report

Step 5:

Open ESF Database Migration Toolkit. Choose **Microsoft SQL Server** (Windows Authentication) as source server. Enter the SQL Server instance name. Username **sa** and default **port 1433** will be used. Enter SQL Server database name **Chinook**.

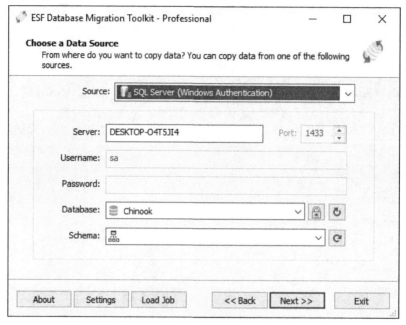

Figure 2-6 Choosing Microsoft SQL Server as source server

Step 6:

Create database **Chinook** on the target MySQL database by running a command:

create database Chinook;

Figure 2-7 Chinook database is created

Step 7:

Choose **MySQL** as a target server. Enter the default server name **localhost** and default port **3306**. Enter username **root** and password. Enter database name **Chinook.**

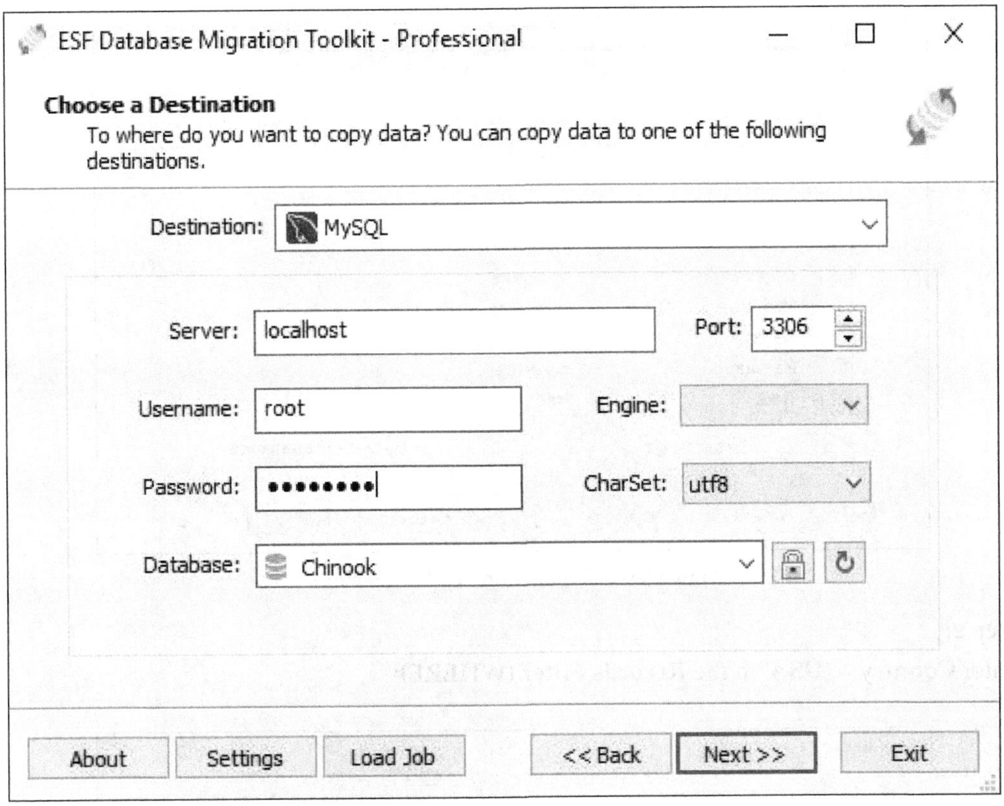

Figure 2-8 Choosing MySQL as a target server

Step 8:

Select tables that you want to migrate from the source to the target. You can change target table name by click a **Destination** filed. Click **... (Transform)** next to **Customer** table and open Transform window:

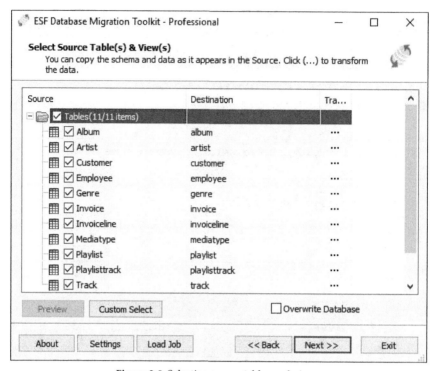

Figure 2-9 Selecting source tables and views

Step 9:

Enter **Country = 'USA'** in the **Records Filter (WHERE)**:

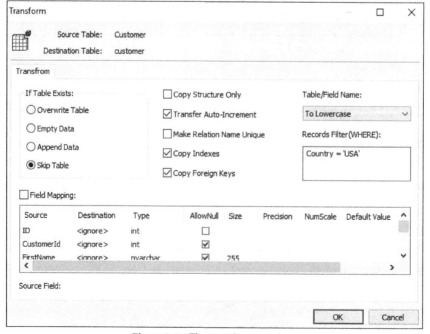

Figure 2-10 The transform window

Step 10:

Click **Next** to see the migration process messages:

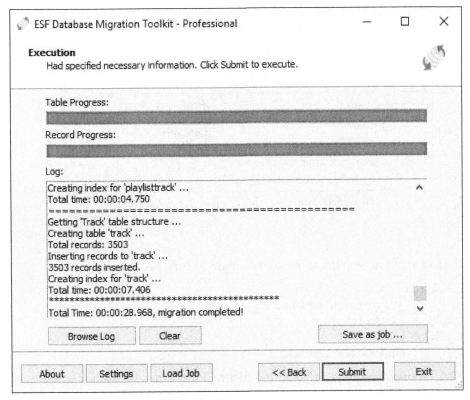

Figure 2-11 The migration messages

Step 11:

Figure 2-11 shows table data types in source and target database. You can see that all the data types matched correctly.

SQL Server Tables	MySQL Tables

⊟ 🔲 dbo.Album
 ⊟ 📁 Columns
   🔑 ID (PK, int, not null)
   📄 AlbumId (int, null)
   📄 Title (nvarchar(255), null)
   📄 ArtistId (int, null)
	Table: album
	
	Columns:
	ID int(11) AI PK
	AlbumId int(11)
	Title varchar(255)
	ArtistId int(11)

⊟ 🔲 dbo.Artist
 ⊟ 📁 Columns
   🔑 ID (PK, int, not null)
   📄 ArtistId (int, null)
   📄 Name (nvarchar(255), null)
	Table: artist
	
	Columns:
	ID int(11) AI PK
	ArtistId int(11)
	Name varchar(255)

Figure 2-12 Table data types in source and target database

Step 12:

Enter the following statements at the query windows then run the query:

```
SELECT
        'playlisttrack' tablename,
        COUNT(*) rows
FROM
        chinook.playlisttrack;
```

Figure 2-13 shows that target MySQL **playlisttrack** table has the same records with the source table (see Figure 2-5):

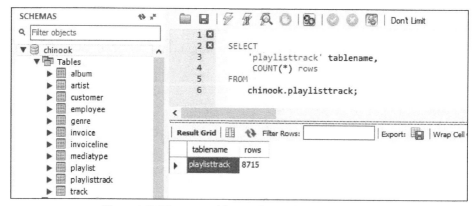

Figure 2-13 The target table records match the source table

Step 13:

Figure 2-14 shows target MySQL **Customer** table. Customers from USA are listed in the **country** field:

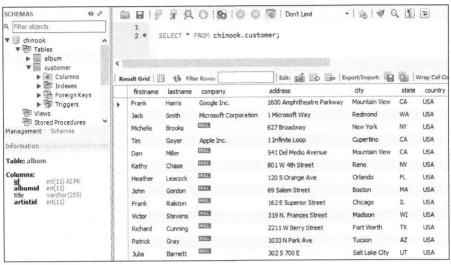

Figure 2-14 The country field in the Customer table

2.1.2 *Using MySQL Workbench Migration Wizard*

Source Server and Database: SQL Server 2016 Test database

Target Server and Database: MySQL 5.7 Test database

Migration Tool: MySQL Workbench Migration Wizard

Steps to migrate SQL Server to MySQL Using MySQL Workbench Migration Wizard:

Step 1:

Using following SQL statements to create a database **Test** and a table **SStoMySQL** in SQL Server 2016:

```
CREATE DATABASE Test
GO

USE [Test]
GO
CREATE TABLE SStoMySQL
( ID BIGINT IDENTITY,
  SAMPLE CHAR(50)
)
INSERT INTO SStoMySQL (SAMPLE)
        VALUES ('Sample Data');
GO 15000 -- Insert 15,000 rows
```

Figure 2-15 Creating a database Test

Step 2:

Open MySQL Workbench and click **Database -> Migration Wizard** in the top menu:

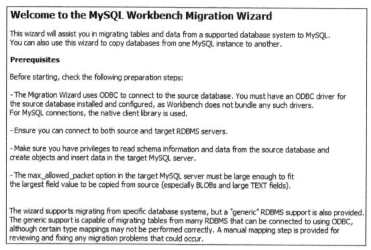

Figure 2-16 Opening MySQL Workbench

Step 3:

Connect source SQL Server database by selecting ODBC (native) and entering server name:

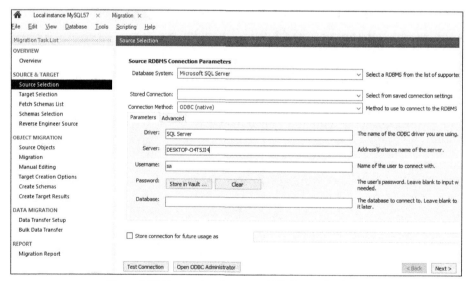

Figure 2-17 Connecting the source SQL Server database

Step 4:

Connect target MySQL server by entering username. Click **Store in Vault...** to enter password.

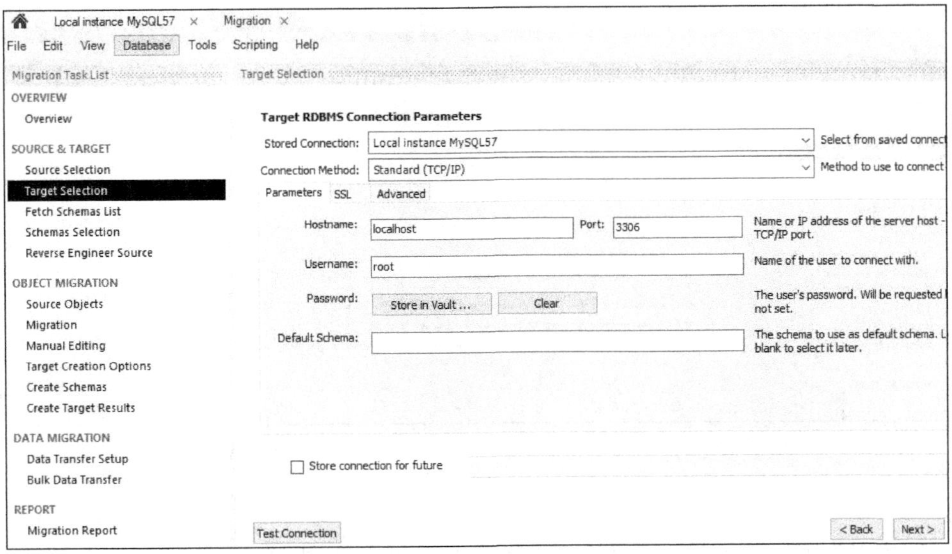

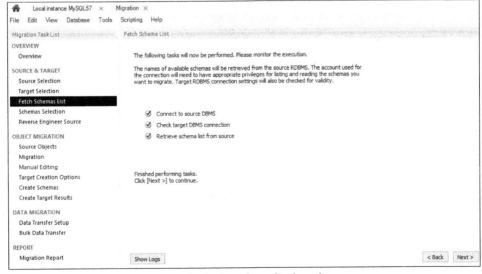

Figure 2-18 Connecting the target MySQL server

Step 5:

Click **Next**. The Migration Wizard will connect the source SQL Server instance and list the databases.

Figure 2-19 Retrieving schema list from the source

Step 6:

Click **Next**. Select the **Test** schema and choose **Only one schema** option.

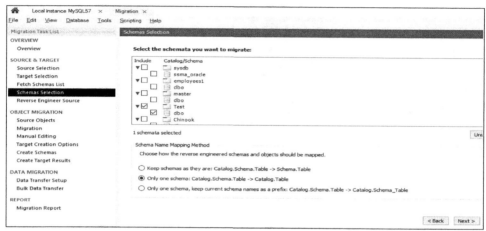

Figure 2-20 Selecting the schemas that you want to migrate

Step 7:

Click **Next**. The Migration Wizard will reverse engineer the **Test** schema.

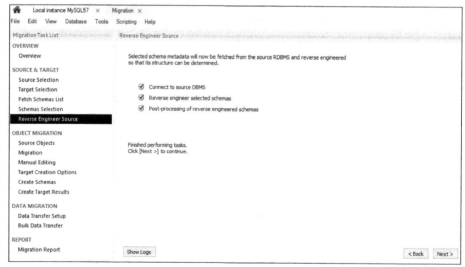

Figure 2-21 Reverse Engineering the selected schemas

Step 8:

Click **Next**. Select the table object(s) to be migrated. Note that Test schema only has one table.

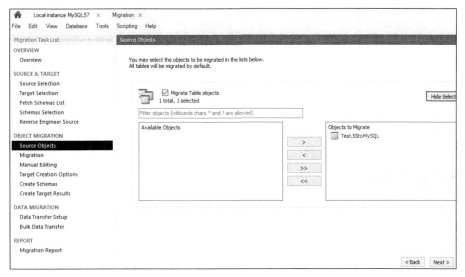

Figure 2-22 Selecting the object(s) to be migrated

Step 9:

Click **Next**. The wizard will convert the SQL Server objects into the equivalent objects in MySQL.

Figure 2-23 Migrating the selected objects

Step 10:

Click **Next**. You can manually edit the migrated tables by clicking **View: Column Mappings**.

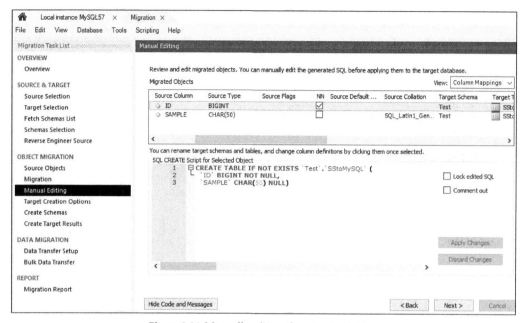

Figure 2-24 Manually editing the migrated tables

Step 11:

Click **Next**. You have option to create a SQL script file here. The wizard creates schema in MySQL.

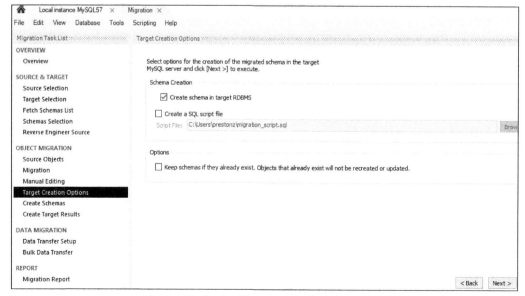

Figure 2-25 Creating the schema in the target server

Step 12:

Click **Next**. The wizard will execute the generated SQL script in the MySQL database.

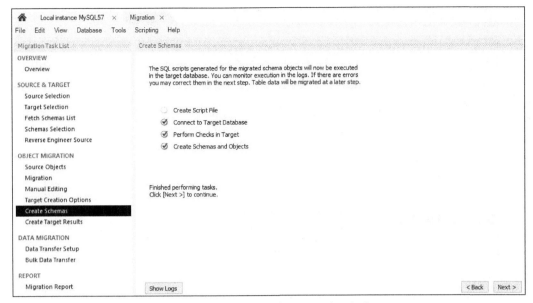

Figure 2-26 The table data will be migrated at the later step

Step 13:

Click **Next**. The wizard shows target Test schema. It moves to data migration stage in the next step.

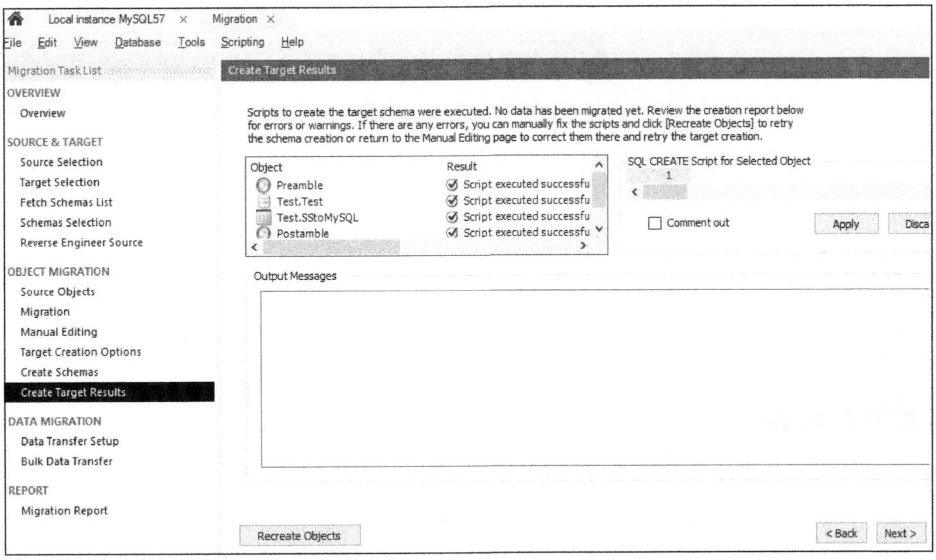

Figure 2-27 Showing the created target schema

Step 14:

Click **Next**. Keep the default data copy method: **Online copy of data to target RDBMS.**

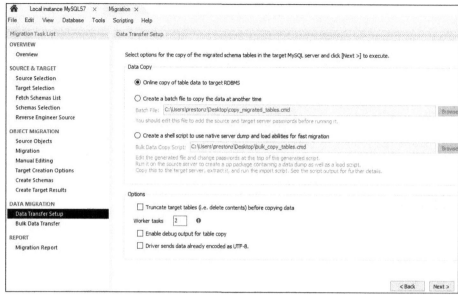

Figure 2-28 Setup migration options

Step 15:

Click **Next**. The data migration steps will take a while if there are many objects to move.

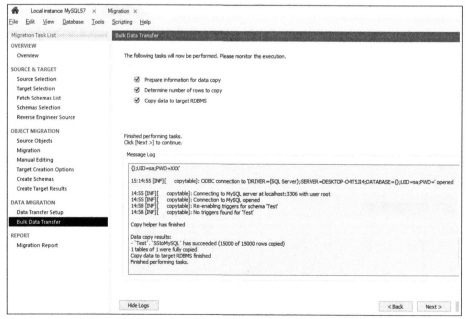

Figure 2-29 Migrating the data to the target server

Step 16:

Click **Next** to close the wizard. You can verify that migration result in target MySQL database.

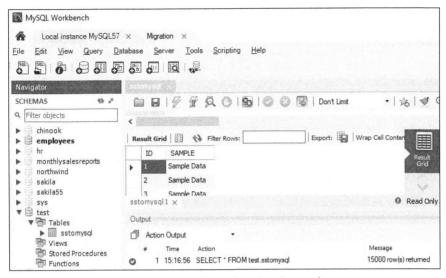

Figure 2-30 Verifying that migration result

2.2 SQL Server to SQL Server Migration Example

Source Server and Database:	SQL Server 2012 Express AdventureWorks2012
Target Server and Database:	SQL Server 2016 AdventureWorks2012
Migration Tool:	Microsoft SQL Server Management Studio

Steps to migrate from SQL Server to SQL Server:

Step 1:

Set up source SQL Server 2012 Express database. Let us download **AdventureWorks2012. bak** from a site below: https://github.com/Microsoft/sql-server-samples/releases/tag/adventureworks

Step 2:

Run the following code from **SQL Server 2012 Express** Query window:

```
use master;

restore database AdventureWorks2012 from disk = N'C:\Temp\AdventureWorks2012.BAK'
WITH File = 1,

Move N'AdventureWorks2012_Data' to N'C:\Temp\AdventureWorks2012_Data.mdf',

Move N'AdventureWorks2012_Log' to N'C:\Temp\AdventureWorks2012_log.ldf',

Nounload, STATS = 10;
```

Figure 2-31 Restoring AdventureWorks2012 database

Step 3:

To see the source **AdventureWorks2012** table records, right-click **AdventureWorks2012 ->**
Reports -> Standard Reports -> Disk Usage by Top Tables:

Table Name	# Records
Sales.SalesOrderDetail	121,317
Production.TransactionHistory	113,443
Production.TransactionHistoryArchive	89,253
Production.WorkOrder	72,591
Production.WorkOrderRouting	67,131
Sales.SalesOrderHeader	31,465
Sales.SalesOrderHeaderSalesReason	27,647
Person.BusinessEntity	20,777
Person.EmailAddress	19,972
Person.Password	19,972
Person.Person	19,972
Person.PersonPhone	19,972
Sales.Customer	19,820
Person.Address	19,614
Person.BusinessEntityAddress	19,614
Sales.CreditCard	19,118

Figure 2-32 AdventureWorks2012 table records

Step 4:

Take look the source **AdventureWorks2012** views from SQL Server Express:

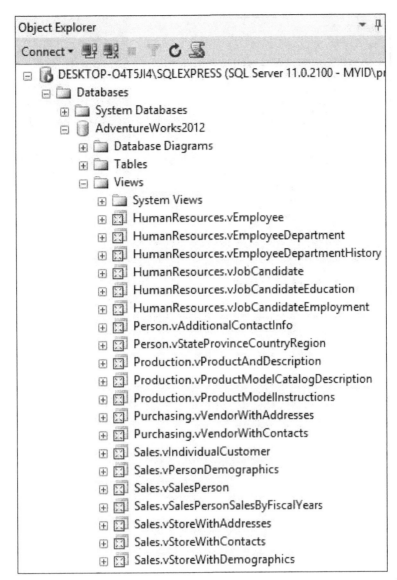

Figure 2-33 Views in AdventureWorks2012

Step 5:

Right-click **AdventureWorks2012** database on SQL Server Express and choose **Tasks -> Back up...**

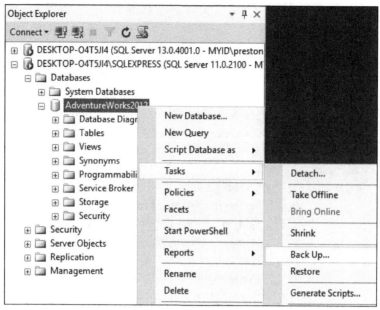

Figure 2-34 AdventureWorks2012 Backup

Step 6:

Add backup location: C:\Temp\AdventureWorks2012_backup:

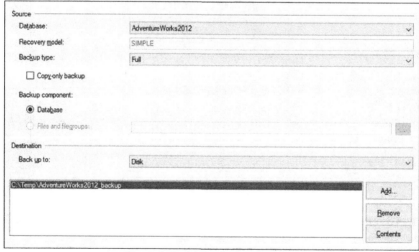

Figure 2-35 Adding backup location

Step 7:

Restore AdventureWorks2012 on SQL Server 2016 by using the backup file from Step 6. Right click **Database** on target SQL Server 2016 and choose **Restore Database...**

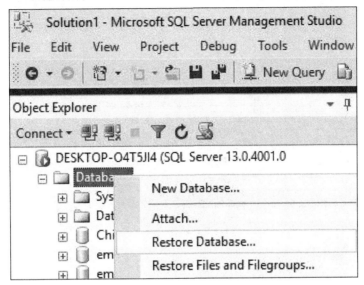

Figure 2-36 Restoring the database

Step 8:

Click **Device** radio button and select the backup file C:\Temp\AdventureWorks2012.bak

Source	
○ Database:	
⦿ Device:	C:\Temp\AdventureWorks2012.bak
Database:	AdventureWorks2012
Destination	
Database:	AdventureWorks2012
Restore to:	The last backup taken (Monday, November 6, 2017 7:51:36 PM)
Restore plan	
Backup sets to restore:	

Restore	Name	Component	Type	Server	Database
☑	AdventureWorks2012-Full Database Backup	Database	Full	BARBKESS24\MSSQL2012RTM	AdventureWorks2012

Figure 2-37 Selecting the backup file

Step 9:

Click OK button. If the restoration is successful you will see a message:

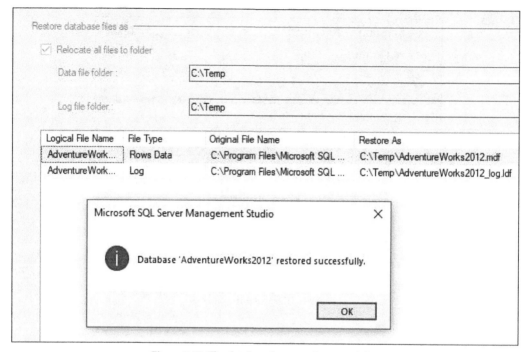

Figure 2-38 The database is restored successfully

Step 10:

After checking the migrated tables and views on target SQL Server 2016 server, we have verified that all the tables, views, stored procedures and functions are migrated from SQL Server 2012 Express to SQL Server 2016.

Figure 2-39 All the tables, views, stored procedures are migrated

Step 11:

SQL Database Studio 3.6.2 Express has a nice feature to display numbers next to the tables, views, stored procedures, functions and triggers. Figure 2-40 shows a comparison for all the objects on the source and the target server.

Figure 2-40 Using SQL Database Studio 3.6.2 Express to display objects numbers

2.3 SQL Server to Oracle Migration Example

Prerequisites:	SQL Server 6.5 and above
	Oracle 9i and above
Source Server and Database:	SQL Server 2016 Chinook database
Target Server and Database:	Oracle 11g Scott database
Migration Tool:	ESF Database Migration Toolkit

Steps to migrate SQL Server to Oracle:

Step 1:

Open ESF Database Migration Toolkit and choose source: **SQL Server (Windows Authentication)**. Enter server name and database name **Chinook:**

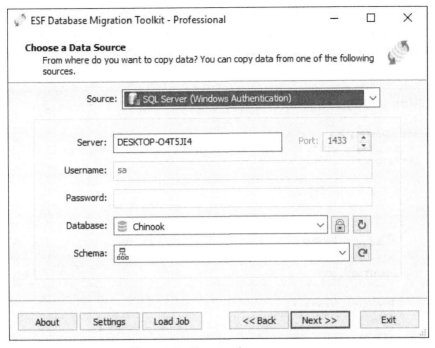

Figure 2-41 Choosing the source server

Step 2:

Setup Oracle target database. Oracle SCOTT schema is used in this chapter. Figure 2-42 displays original tables in SCOTT schema. Drop the four tables.

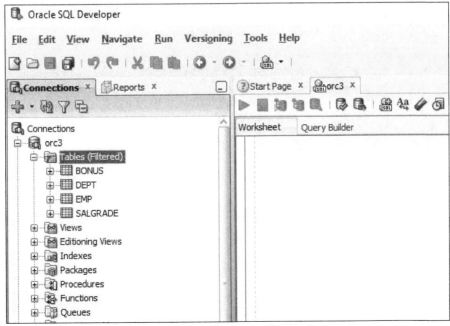

Figure 2-42 Original tables in SCOTT schema

Step 3:

Choose Oracle as target database. Enter server **localhost,** user name **scott** and password. Enter database SID orcl3:

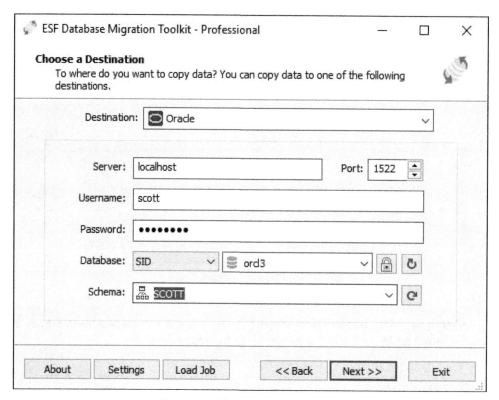

Figure 2-43 Choosing the target server

Step 4:

Select tables that you want to migrate from the source to the target:

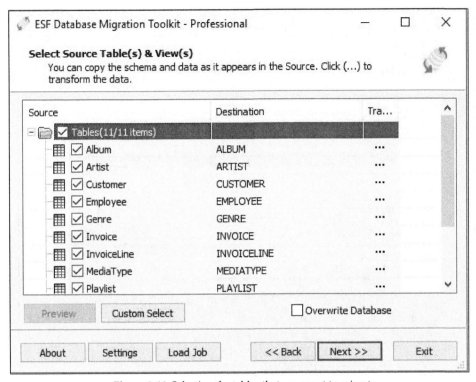

Figure 2-44 Selecting the tables that you want to migrate

Step 5:

Click **Next** and then click **Submit** button:

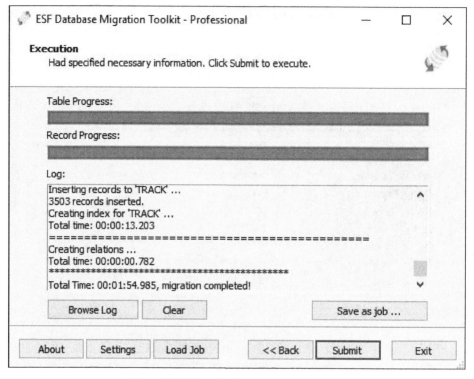

Figure 2-45 The migration progress messages

Step 6:

Go to Oracle SQL Developer and connect to SCOTT schema:

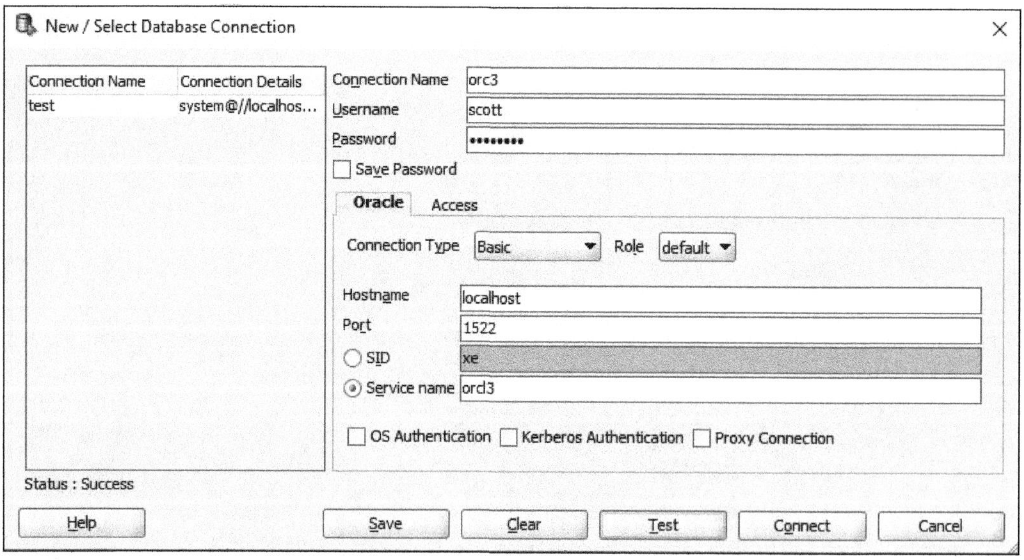

Figure 2-46 Connecting to SCOTT schema

Step 7:

Enter the following SQL code in query window and run the query to show all the table records:

```
SELECT
    table_name, num_rows
FROM dbo.tables
WHERE owner = 'SCOTT'
ORDER By num_rows DESC;
```

Figure 2-47 Oracle SQL code that shows all the table records

Step 8:

We have verified that all the tables and data are migrated from SQL Server to Oracle. You can see that although Oracle SQL Server has a data migration tool available, it's not as smooth as ESF database migration toolkit.

	TABLE_NAME		NUM_ROWS
1	PLAYLISTTRACK		8715
2	TRACK		3503
3	INVOICELINE		2240
4	INVOICE		412
5	ALBUM		347
6	ARTIST		275
7	CUSTOMER		59
8	GENRE		25
9	PLAYLIST		18
10	EMPLOYEE		8

Figure 2-48 Oracle table records after the migration

Figure 2-49 shows data types in Album and Invoice tables in source and target database:

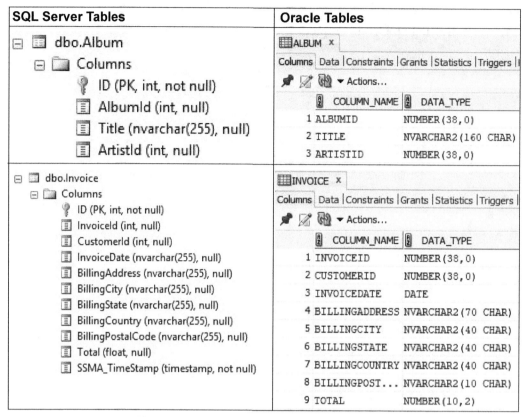

Figure 2-49 Data types comparison for the two servers

Summary

Chapter 2 covers the following:

- Using ESF database migration toolkit and MySQL Migration Wizard to migrate from SQL Server to MySQL. The ESF database migration toolkit takes four steps and MySQL Migration Wizard needs 16 steps.
- Using SQL Server Management Studio to migrate from SQL Server to SQL Server
- Using ESF database migration toolkit to migrate from SQL Server to Oracle

Chapter 3

More About SQL Server Management Studio

We have used SQL Server Management Studio in the last chapter, I would like to introduce other useful features in the studio: Visual Database Design, Query Designer, Generating Scripts, Activity Monitor, Query Options, Template Browser, Database Engine Tuning Advisor, Disk Usage Report, Database Properties and Backup.

3.1 Visual Database Design

SQL Server Management Studio provides a useful tool for developers and DBAs to create database diagrams.

Steps to create a relational diagram:

Step 1:

Create a database **Course** and a table **Courses** with three fields (courseID, Title and Credits):

```
☐ 🛢 Course
   ☐ 📁 Database Diagrams
   ☐ 📁 Tables
      ☐ 📁 System Tables
      ☐ 📁 FileTables
      ☐ 🗐 dbo.Courses
         ☐ 📁 Columns
            🔑 courseID (PK, int, not null)
            🗐 Title (varchar(50), not null)
            🗐 Credits (int, not null)
```

Figure 3-1 Course database and Courses table are created in SQL Server

Step 2:

Create a table **Instructors** with four fields (instructorID, FirstName, LastName and HireDate):

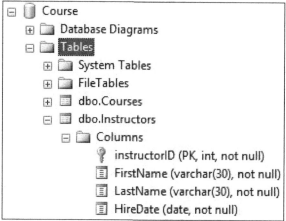

Figure 3-2 The Instructor table is created in SQL Server

Step 3:

Create a table **Course_Instructor** with two fields (instructorID, courseID):

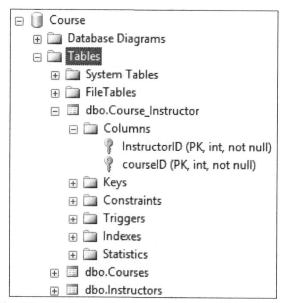

Figure 3-3 The Course_Instructor table is created in SQL Server

Step 4:

Right-click **Database Diagrams** to open a database diagram screen. Add the three tables to the diagram:

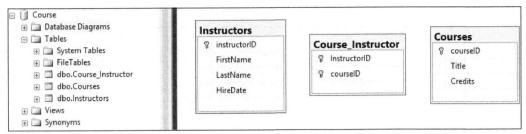

Figure 3-4 The Instructor table is created in SQL Server

Step 5:

Drag the **instructorID** on the Instructor table to the **instructorID** on the Course_Instructor table, a window appears with a foreign key name and related fields from the two tables. Click **OK** button.

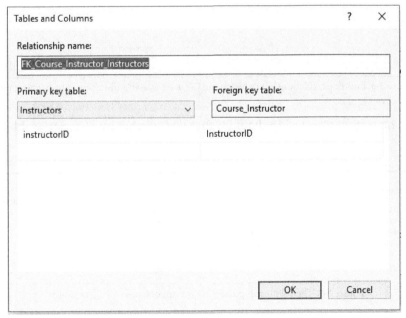

Figure 3-5 Building a relationship between the Instructor table and the Course_Instructor table

Step 6:

Drag the **courseID** on the Course table to the **courseID** on the Course_Instructor table, a window appears with a foreign key name and linked fields from the two tables. Click **OK** button.

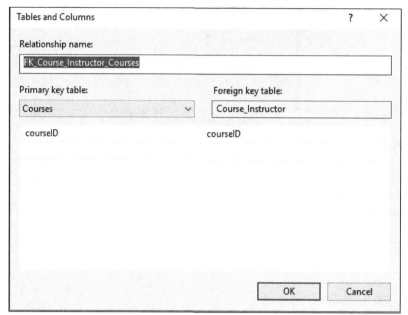

Figure 3-6 Building a relationship between the Course table and the Course_Instructor table

Figure 3-7 shows the relationship diagram that was created by SQL Server Management Studio:

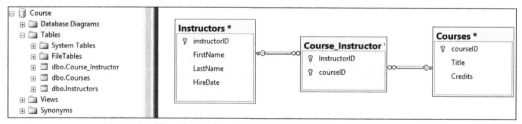

Figure 3-7 The relationship diagram is built between the three tables

3.2 SQL Server Query Designer

Query Designer is a visual tool to help developers and DBAs to pull query together.
To open **Query Designer** window, click **Query -> Design Query in Editor:**

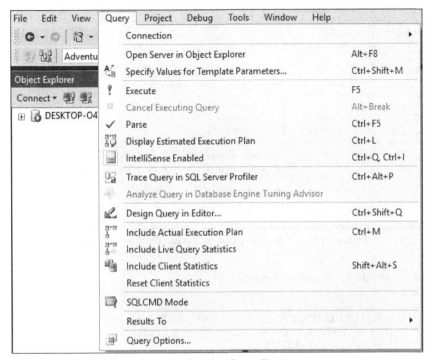

Figure 3-8 Opening Query Designer

Query designer is a very useful tool especially when you work on joins in queries. For example, we want to write a query to display the country name, city, and the departments which are in the country. We will use HR database that was migrated to SQL Server from Oracle in the Chapter 2. Add the **Department**, **Locations** and **Countries** tables to Query Designer. Select **Department_Name**, **City**, **State_Province** and **Country_Name** fields then the SQL JOIN statement shows at the bottom section of the Query Designer.

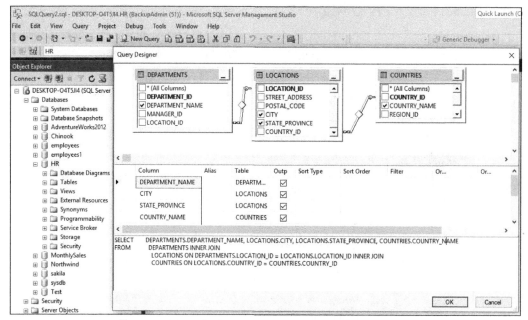

Figure 3-9 Using the Query Designer to join the three databases

Below is the generated SQL join statement:

```
SELECT DEPARTMENTS.DEPARTMENT_NAME, LOCATIONS.CITY,
LOCATIONS.STATE_PROVINCE, COUNTRIES.COUNTRY_NAME
  FROM      DEPARTMENTS
     INNER JOIN
                LOCATIONS ON
        DEPARTMENTS.LOCATION_ID = LOCATIONS.LOCATION_ID
     INNER JOIN
                COUNTRIES ON
                LOCATIONS.COUNTRY_ID = COUNTRIES.COUNTRY_ID
```

Figure 3-10 SQL JOIN statement

Click on **OK** button and **Execute** the query to get the result below:

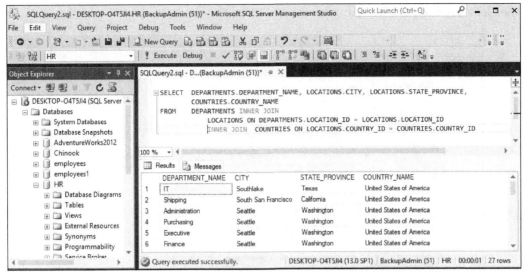

Figure 3-11 Getting the query result

3.3 Generating Scripts

It is a very useful feature in SQL Server Management Studio especially when moving objects between SQL Server instances or migrating to other database systems. Below are the steps to generate a script:

Step 1:

Right-clicking the database and select **Tasks -> Generating Scripts**:

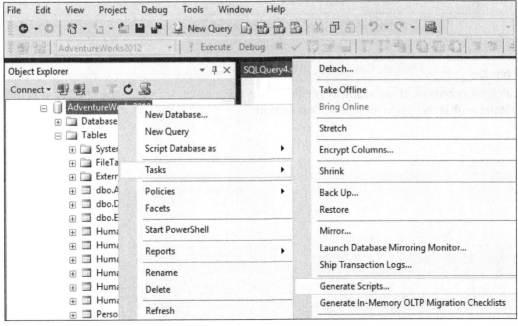

Figure 3-12 Opening Generate Scripts wizard

The **Generate Scripts** wizard appears:

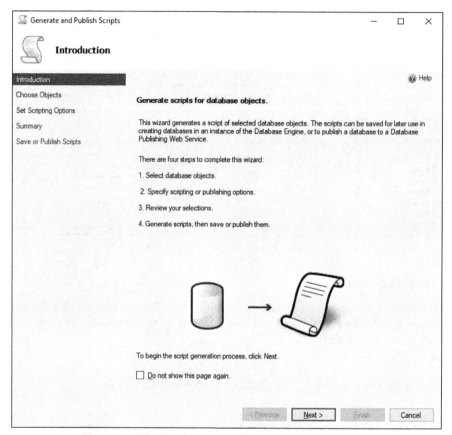

Figure 3-13 The introduction page for Generate Scripts wizard

Step 2:

Click Next button. If we want to get the script for a function, we choose **Select specific database objects** option and click **Next** button:

Figure 3-14 Selecting the database objects to script

Step 3:
Select the output type and save the script to a location, for example, C:\temp\script.sql:

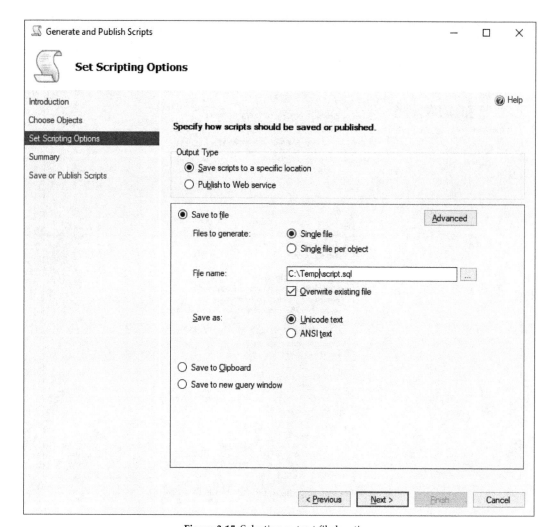

Figure 3-15 Selecting output file location

Step 4:

Click the **Next** button. You will see the following screen:

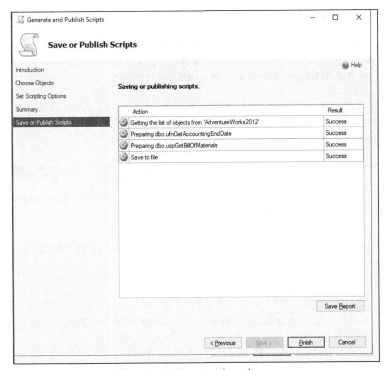

Figure 3-16 The wizard result page

Step 5:

Click the **Finish** button to see the generated script:

```
  8
  9    CREATE FUNCTION [dbo].[ufnGetAccountingEndDate]()
 10    RETURNS [datetime]
 11    AS
 12    BEGIN
 13        RETURN DATEADD(millisecond, -2, CONVERT(datetime, '20040701', 112));
 14    END;
 15
 16    GO
 17    /****** Object:  StoredProcedure [dbo].[uspGetBillOfMaterials]      ******/
 18    SET ANSI_NULLS ON
 19    GO
 20    SET QUOTED_IDENTIFIER ON
 21    GO
 22
 23    CREATE PROCEDURE [dbo].[uspGetBillOfMaterials]
 24        @StartProductID [int],
 25        @CheckDate [datetime]
 26    AS
 27    BEGIN
 28        SET NOCOUNT ON;
 29
 30        -- Use recursive query to generate a multi-level Bill of Material (i.e. all level 1
 31        -- components of a level 0 assembly, all level 2 components of a level 1 assembly)
 32        -- The CheckDate eliminates any components that are no longer used in the product on this date.
 33        WITH [BOM_cte]([ProductAssemblyID], [ComponentID], [ComponentDesc], [PerAssemblyQty], [StandardCost],
 34            [ListPrice], [BOMLevel], [RecursionLevel]) -- CTE name and columns
 35        AS (
 36            SELECT b.[ProductAssemblyID], b.[ComponentID], p.[Name], b.[PerAssemblyQty], p.[StandardCost],
 37                p.[ListPrice], b.[BOMLevel], 0 -- Get the initial list of components for the bike assembly
 38            FROM [Production].[BillOfMaterials] b
 39                INNER JOIN [Production].[Product] p
 40                ON b.[ComponentID] = p.[ProductID]
 41            WHERE b.[ProductAssemblyID] = @StartProductID
 42                AND @CheckDate >= b.[StartDate]
 43                AND @CheckDate <= ISNULL(b.[EndDate], @CheckDate)
 44            UNION ALL
```

Figure 3-17 The generated script

3.4 Activity Monitor

SQL Server Activity Monitor displays information about the SQL Server instance. The users can see the SQL Server performance through graphical overview. To open Activity Monitor, click on the Activity Monitor icon from the toolbar:

Figure 3-18 Activity Monitor icon

Or you can right click the SQL Server instance and select **Activity Monitor**, an overview window opens:

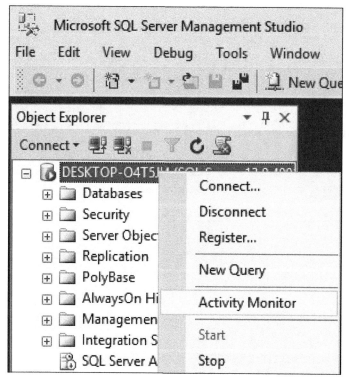

Figure 3-19 Activity Monitor option

The overview section displays the processor time (%), the waiting tasks, the data file input/output rate and the batch requests.

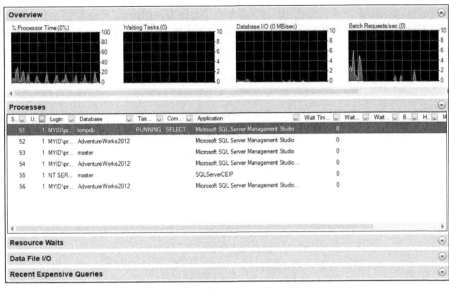

Figure 3-20 The overview screen

3.5 Query Options

To open **Query Options** window, click **Query** -> **Query Options** on the top menu:

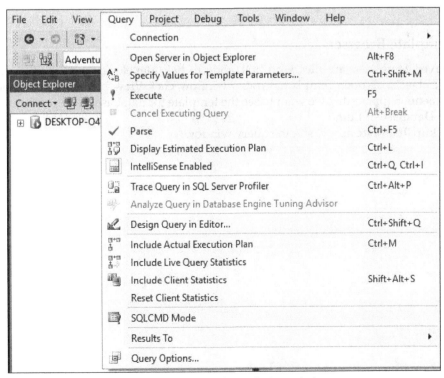

Figure 3-21 Opening Query Options window

Click **Results** -> **Grid** and check the checkbox **Include column headers when copying or saving the results.** When you export query results to user(s) the column headers are usually required.

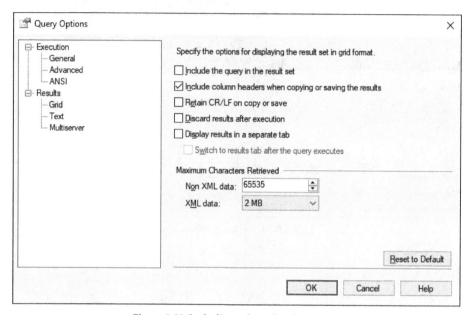

Figure 3-22 Including column header option

3.6 Template Browser

SQL Server Templates are files with SQL code to help developers and DBAs to create database objects. To open Template Browser window, click **View** -> **Template Browser** on the top menu. Suppose that we want to see the template for database backup, we can select **Backup Database** -> **Edit**:
The **Backup Template** appears in the query window:

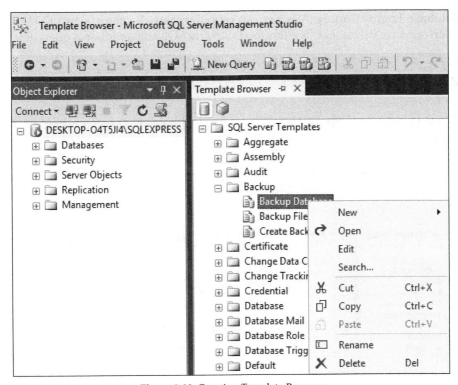

Figure 3-23 Opening Template Browser

3.7 Database Engine Tuning Advisor

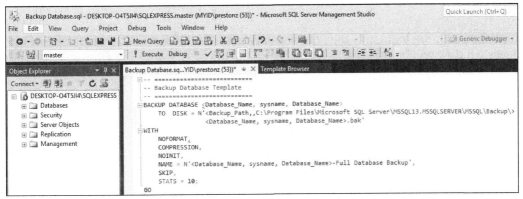

Figure 3-24 Backup Database Template

The **Database Engine Tuning Advisor** can help database developers and DBAs to find query performance issues. Let us enter the following SQL statement on a query window:

SELECT D.department_name, L.city, L.state_province

FROM departments D JOIN locations L ON D.location_id = L.location_id;

Right-click the query window to open **Analyze Query in Database Engine Tuning Advisor**:

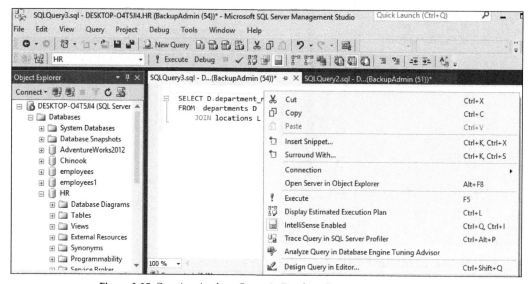

Figure 3-25 Opening Analyze Query in Database Engine Tuning Advisor

A session with the login and the date was created. The database HR that is used in the query window will be selected.

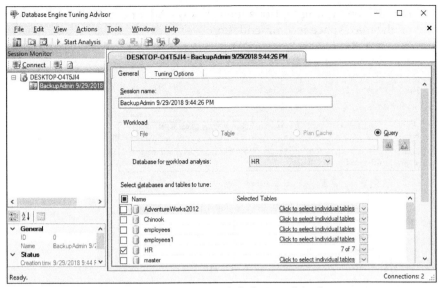

Figure 3-26 A session with the login and the date

Click on **Start Analysis** button. The analysis process will take a while. If there is an error the process will be stopped with a message:

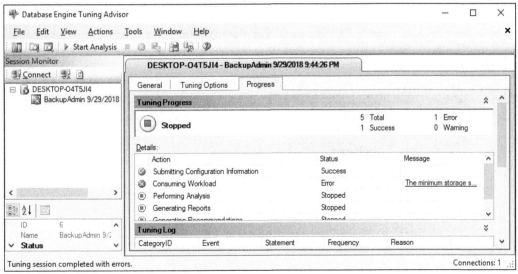

Figure 3-27 Analysis progress window

To fix the error click on **Tuning Options** then click on **Advanced Options** button, enter 5 (MB) based on the max. space recommendation:

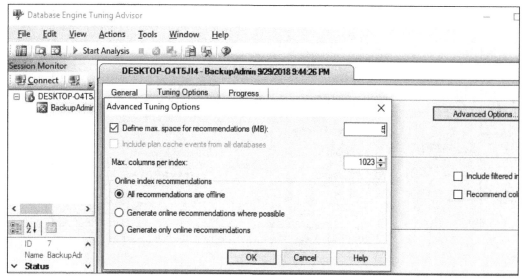

Figure 3-28 Advanced Tuning Options

Click on **OK** button and click on **Start Analysis** again:

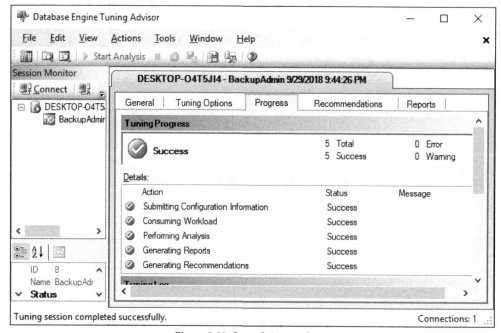

Figure 3-29 Second time analysis

3.8 Disk Usage Report

Right-click a database and select **Reports** -> **Standard Reports** -> **Disk Usage** to get a Disk Usage Report:

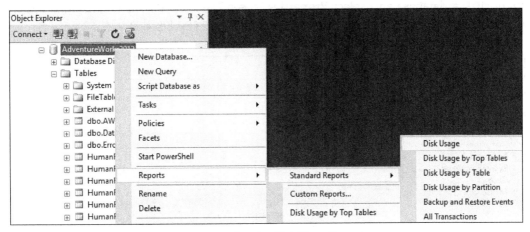

Figure 3-30 Opening Disk Usage Report

Below is a sample **Disk Usage** Reports:

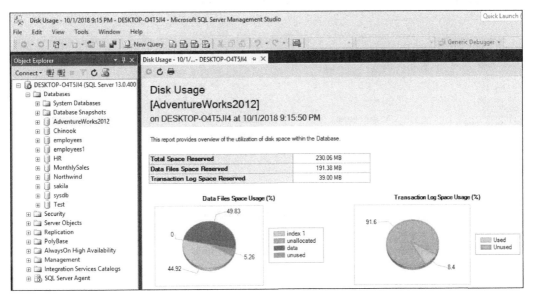

Figure 3-31 Disk Usage Reports for AdventureWorks2012

3.9 Database Properties

Database properties page allows you modify database initial size and autogrowth rate by MB or by percentage. You can also change database recovery model from the Options page.

Right-click a database and select **Properties** to see Database Properties window. Select **Files** on the left pane you will see the initial data file size and log size:

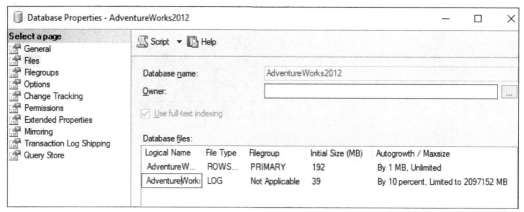

Figure 3-32 Opening Database Properties page

Select **Options** on the left pane to see the recovery model:

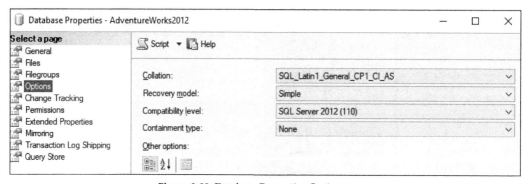

Figure 3-33 Database Properties Options page

3.10 SQL Server Backup

SQL Server best practice suggests that backups are stored on the different location with database files. For demo purpose the backup files and the database files are saved at C drive in this book.

3.10.1 Using T-SQL

White the following T-SQL in the query window and click **Execute** query button. You will see that the backup is successfully done at the C drive:

> BACKUP DATABASE [Chinook]
>
> TO DISK = N'C:\MSSQLBackup\Chinook.bak'

Figure 3-34 shows the backup file name, type, datetime and size:

Figure 3-34 T-SQL backup command

3.10.2 Using Agent Job

Steps to schedule a backup using SQL Server Agent:

Step 1: Right-click the **Agent** and select New Job.

Step 2: Enter the job name Backup and click **OK** button.

Step 3: Right-click the Backup job and select **Property**.

Step 4: Click **Steps** on the left pane then click **New** to create a new step.

Step 5: Enter the following SQL script in Notepad:

```
DECLARE @name VARCHAR(50) -- database name
DECLARE @path VARCHAR(256) -- path for backup files
DECLARE @fileName VARCHAR(256) -- filename for backup
DECLARE @fileDate VARCHAR(20) -- used for file name


SET @path = 'C:\MSSQL\Full_Backup\'

SELECT @fileDate = CONVERT(VARCHAR(20),GETDATE(),112)

DECLARE db_cursor CURSOR FOR

SELECT name
FROM master.dbo.sysdatabases
WHERE name NOT IN ('master','model','msdb','tempdb')

OPEN db_cursor
FETCH NEXT FROM db_cursor INTO @name

WHILE @@FETCH_STATUS = 0
BEGIN
    SET @fileName = @path + @name + '_' + @fileDate + '.BAK'
    BACKUP DATABASE @name TO DISK = @fileName

    FETCH NEXT FROM db_cursor INTO @name
END

CLOSE db_cursor
DEALLOCATE db_cursor
```

Figure 3-35 T-SQL backup script

Step 6: Enter step name Step 1 and paste the SQL script in the command pane:

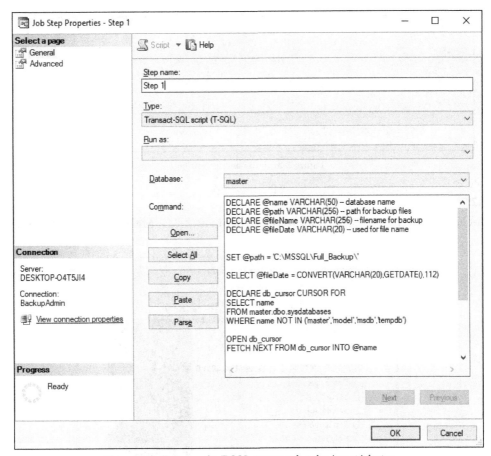

Figure 3-36 Entering the T-SQL command at the Agent job step

Step 7: Click **OK** button.

Step 8: Click **Schedules** on the left pane and click **New** button to setup a schedule.

3.10.3 *Using Maintenance Plan*

SQL Server Database Maintenance Plans can help DBAs to do a lot of database administration tasks. For example, database backup, checking database integrity, cleaning old backup files, etc.

Steps to do log backup:

Step 1:

Create a full backup. You can use T-SQL or maintenance plan for a full backup. Below is a full back for three databases:

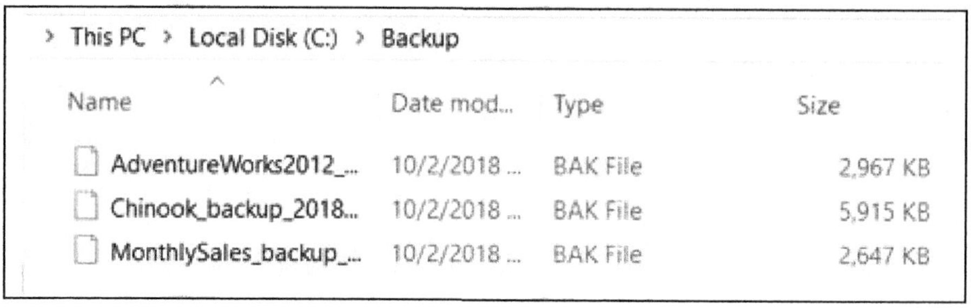

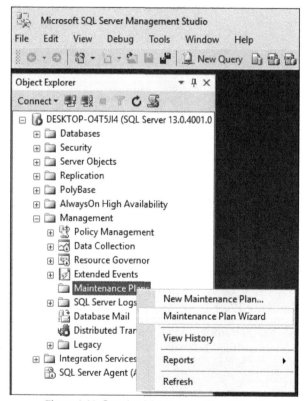

Figure 3-37 Full backup results

Step 2:

Right-click **Maintenance Plan** under **Management** and select **Maintenance Plan Wizard**:

Figure 3-38 Opening Maintenance Plan Wizard

Step 3:

SQL Server Maintenance Plan Wizard appears:

Figure 3-39 Maintenance Plan Wizard starting page

Step 4:

Enter the plan name and click **Change** button to set up a schedule:

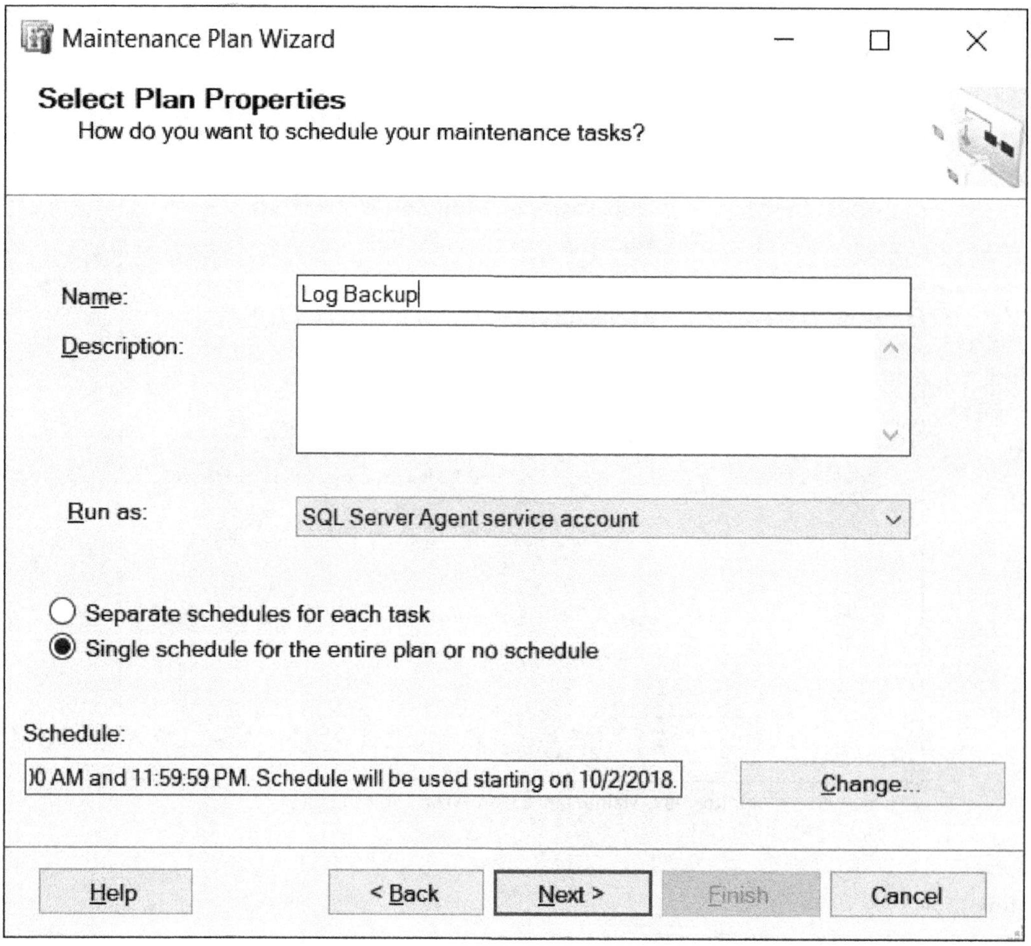

Figure 3-40 Setup a backup schedule

Step 5:

Select **Back Up Database (Transaction Log)** option:

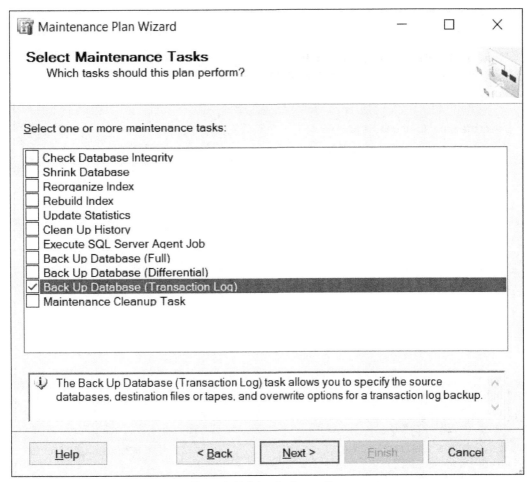

Figure 3-41 Select backup options

Step 6:

Order the tasks If you select several tasks:

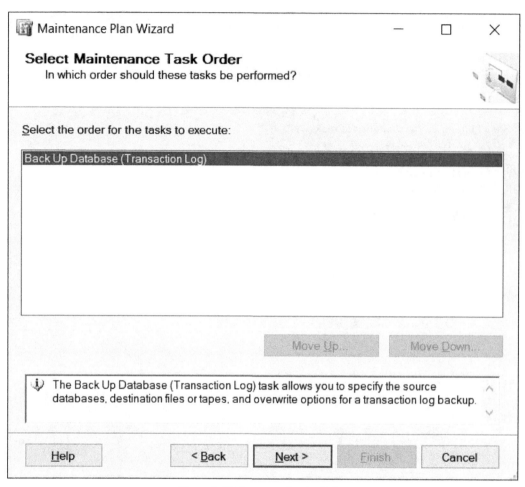

Figure 3-42 You can order the tasks

Step 7:

Click on **Database(s)** drop-down list to select **all user databases** and click **OK** button:

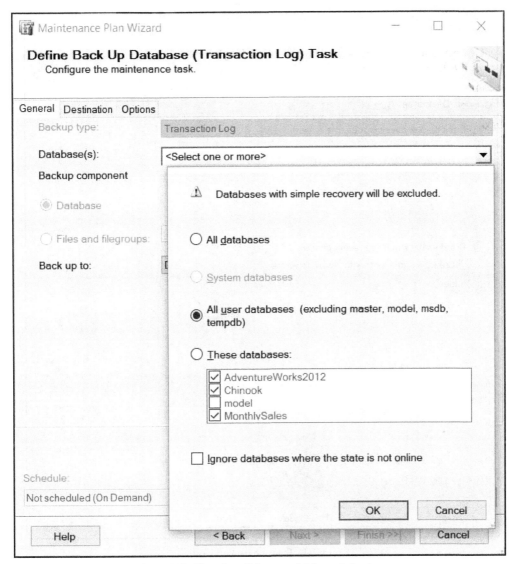

Figure 3-43 Choosing all the user databases to backup

Step 8:

Click **Destination** tab to save the log files to a specific location. For example, C:\Backup folder. Select **Create a backup file for every database** option. You can change the backup file extension. For example, **TRN** extension is used here instead of the default **trn** extension.

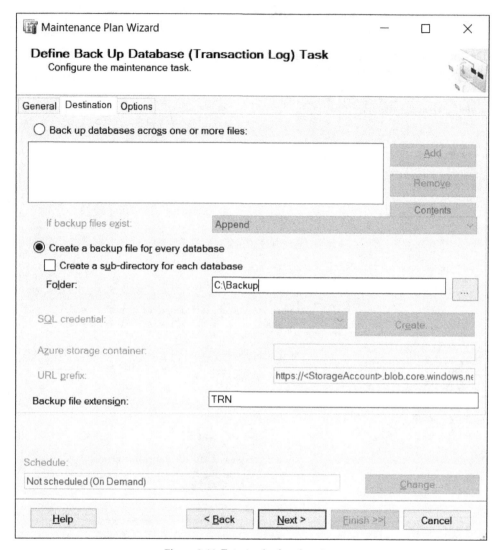

Figure 3-44 Entering backup location

Step 9:

You can change the report location at the screen below:

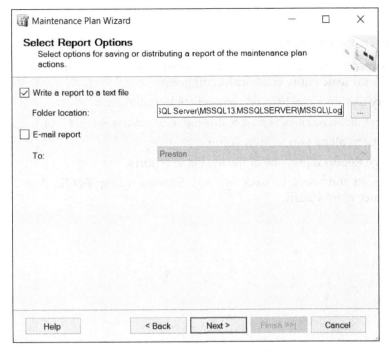

Figure 3-45 You can change the report location

Step 10:

Click **Next** to see the Maintenance Plan Wizard progress:

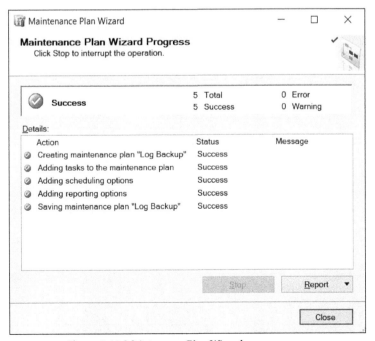

Figure 3-46 Maintenance Plan Wizard progress page

Summary

Chapter 3 covers the following:

- SQL Server Management Studio Visual Database Design can help database developers and DBAs to create entity relationship diagrams.
- The Query Designer can help you to write SQL statements.
- You can get database objects scripts through Generating Scripts tool.
- Query options allow you to view column headers.
- Disk Usage Report is just one of the available reports.
- There are several ways to back up SQL Servers: using T-SQL, Agent jobs or the Maintenance Plan wizard.

Chapter 4

MySQL Database Migration

The main topics in this chapter are illustrated in Figure 4-1: MySQL database migration to MySQL, MySQL database migration to SQL Server and MySQL database migration to Oracle.

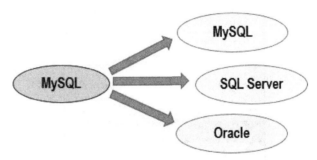

Figure 4-1 MySQL migration paths

4.1 MySQL to MySQL Migration Example

Source Server and Database: MySQL 5.5 Sakila database
Target Server and Database: MySQL 5.7 Sakila55 database
Migration Tool: MySQL SQL export commands and Import Wizard

4.1.1 Using MySQL Export Commands and Import Wizard

Steps to migrate from MySQL to MySQL using MySQL export commands and Import wizard:

Step 1:

The first method is using MySQL Import Wizard. First, let us prepare a source MySQL server as an example. Download archived MySQL 5.5 from https://download.mysql.com

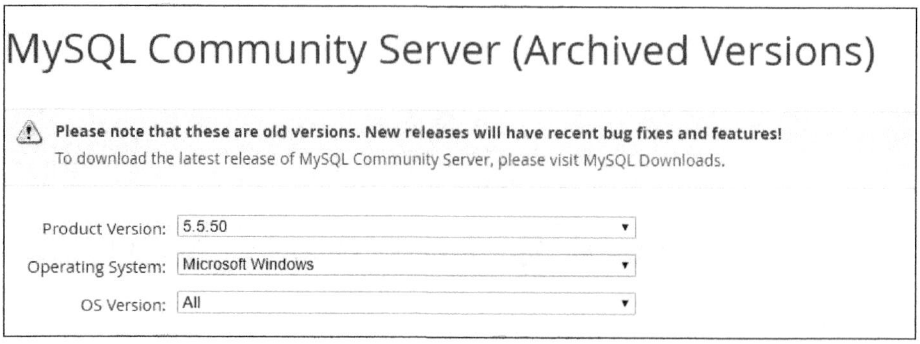

MySQL Community Server (Archived Versions)

⚠ **Please note that these are old versions. New releases will have recent bug fixes and features!**
To download the latest release of MySQL Community Server, please visit MySQL Downloads.

Product Version: | 5.5.50 ▼
Operating System: | Microsoft Windows ▼
OS Version: | All ▼

Figure 4-2 MySQL installer archived versions

Step 2:

After MySQL 5.5 installation, sakila database tables, views and routines appear in the Object Browser:

Figure 4-3 Sakila database structures

Step 3:

Open Command Prompt and enter the following command to export the source sakila database to sakila55.sql file. Please note that if you do not include **"-routines"** in the command all the views and routines will not be exported.

> *C:\Program Files\MySQL Server 5.5\bin > mysqldump.exe -e **–routines**
> -uroot -ppassword sakila > F:\Test\sakila55.sql*

Figure 4-4 Export command for sakila database

Step 4:

You can also export schema(s), stored procedures and functions using MySQL Workbench: Right click **Data export** in the Management section and select the schema sakila. Check **Dump Stored procedures and Functions** then click **Start Export**:

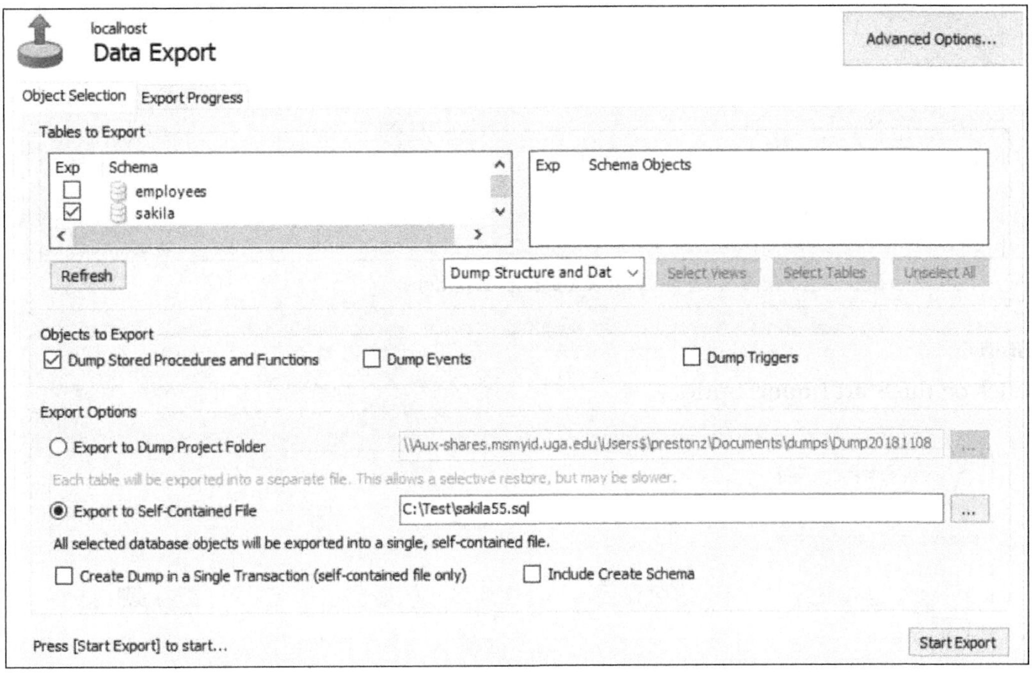

Figure 4-5 MySQL Data Export Wizard

Step 5:

Now Import sakila55 to target MySQL 5.7 server by clicking **Data Import** tool in the left pane. Select **Import from Self-Contained File** and select the backup file *sakila55.sql*.

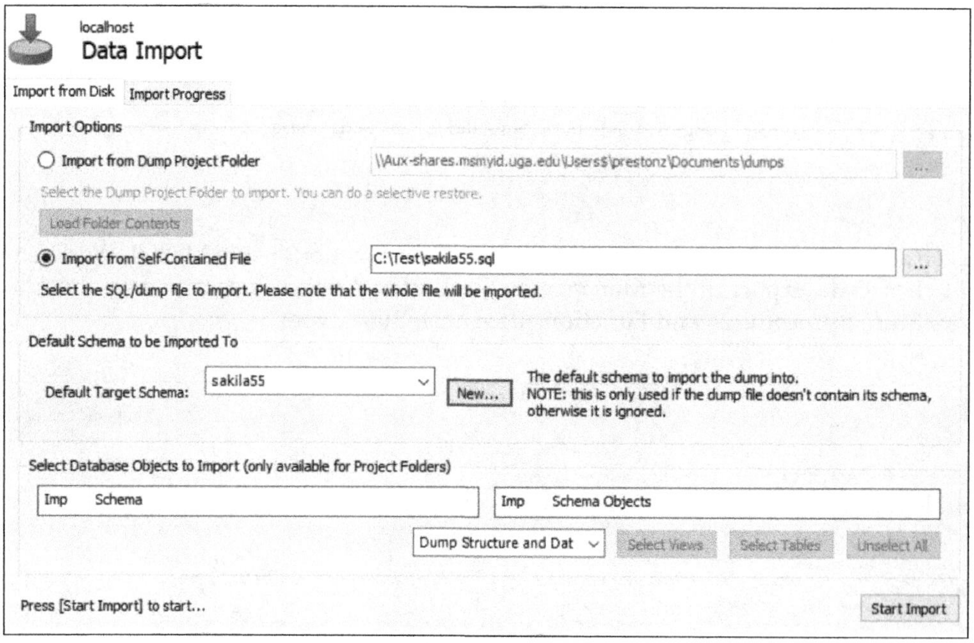

Figure 4-6 Import options

Step 6:

Click on the **Start Import** button.

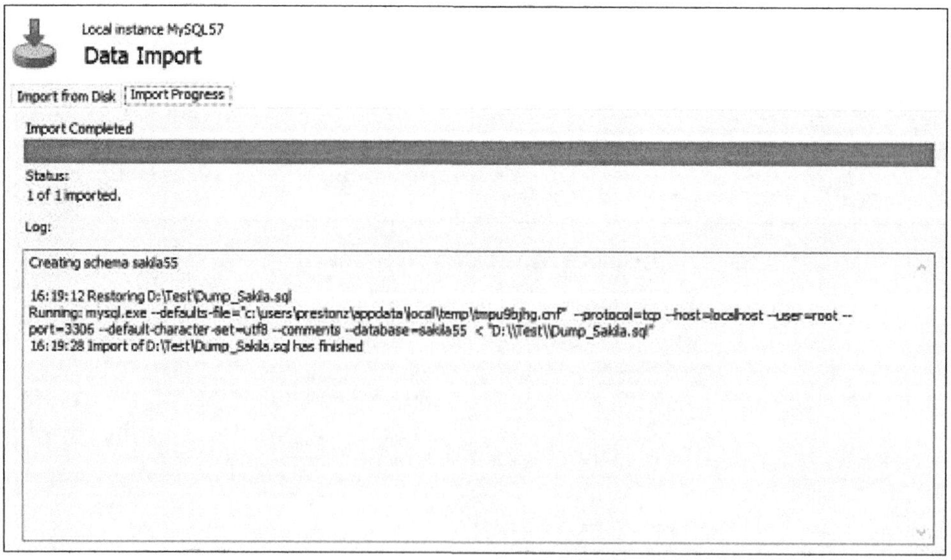

Figure 4-7 Data Import status

Below is the source Sakila55.payment table records: 16,049 rows.

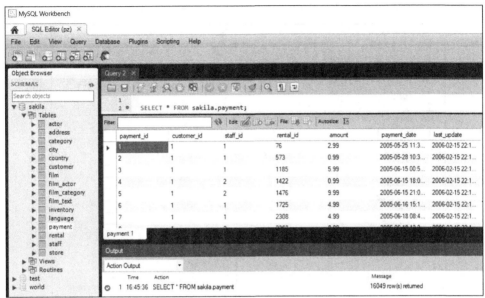

Figure 4-8 Source payment table records

Figure 4-9 shows target Sakila55.payment table records: 16,049 rows.

payment_id	customer_id	staff_id	rental_id	amount	payment_date	last_update
1	1	1	76	2.99	2005-05-25 11:30:37	2006-02-15 22:12:30
2	1	1	573	0.99	2005-05-28 10:35:23	2006-02-15 22:12:30
3	1	1	1185	5.99	2005-06-15 00:54:12	2006-02-15 22:12:30
4	1	2	1422	0.99	2005-06-15 18:02:53	2006-02-15 22:12:30

payment 1 ×

Output

Action Output

#	Time	Action	Message
1	08:53:57	SELECT * FROM sakila55.payment	16049 row(s) returned

Figure 4-9 Target payment table records

Figure 4-10 shows that all the stored views, stored procedures and functions are migrated from MySQL 5.5 to MySQL 5.7.

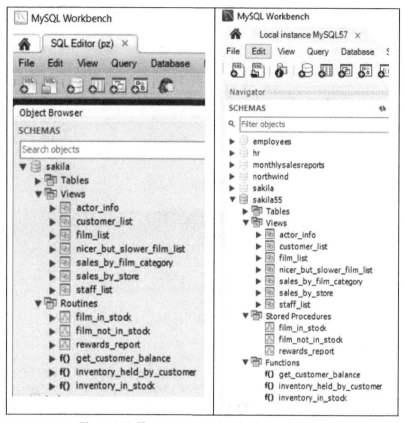

Figure 4-10 The comparison after database migration

4.1.2 Using MySQL Workbench Schema Transfer Wizard

Source Server and Database:	MySQL 5.5 Sakila database
Target Server and Database:	MySQL 5.7 Sakila database
Migration Tool:	MySQL Workbench Schema Transfer Wizard

Steps to migrate from MySQL to MySQL using MySQL Workbench Schema Transfer Wizard:

Step 1:

The second method is using MySQL Workbench Schema Transfer Wizard. It can be used to upgrade MySQL database. Open MySQL Workbench and select **Database ->Schema Transfer Wizard** from the top menu:

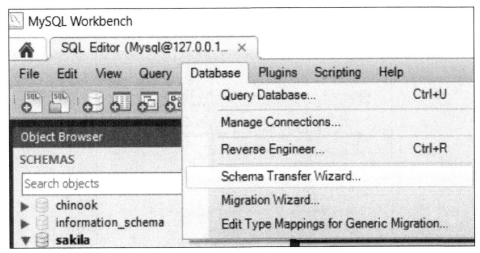

Figure 4-11 Opening Schema Transfer Wizard

MySQL Workbench Schema Transfer Wizard page appears: The message said that the tool is for developer machines.

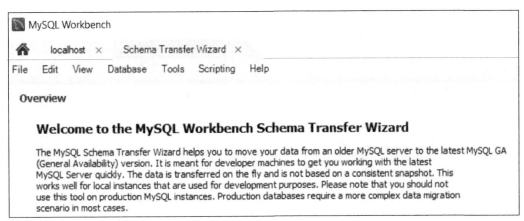

Figure 4-12 Schema Transfer Wizard welcome page

Step 2:

Click **Start the Wizard**. Select source MySQL instance (5.5) and target MySQL instance (5.7). Test connections then click Next.

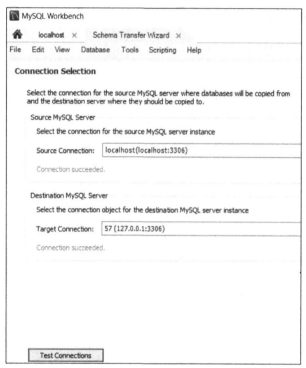

Figure 4-13 Selecting source and target connection

Step 3:

Choose the schema sakila then and click **Start Copy** button:

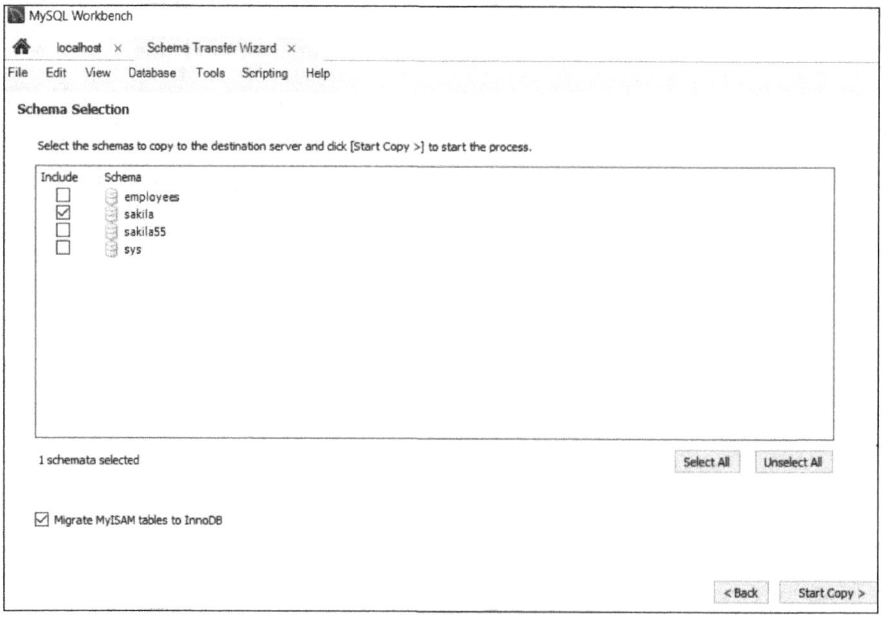

Figure 4-14 Selecting sakila schema

Step 4:

The log messages show that 15 tables of 16 were successfully migrated.

Figure 4-15 The migration process status

Step 5:

Click **Next** to see the migration results:

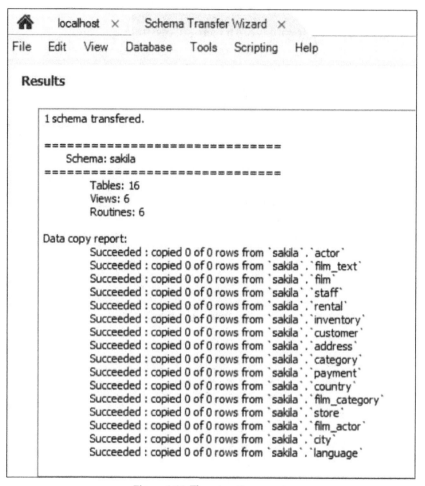

Figure 4-16 The migration report

Figure 4-17 shows the source and target database after the migration.

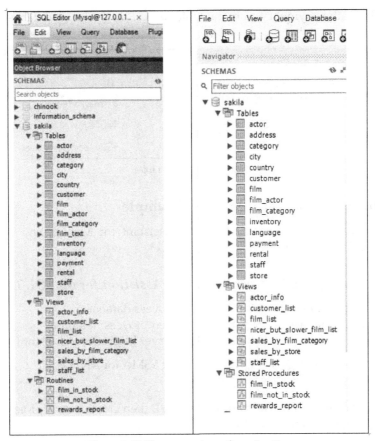

Figure 4-17 The comparison after migration

Figure 4-18 shows the payment and rental table rows from the source database:

	Time	Action	Message
1	16:32:59	USE sakila	0 row(s) affected
2	16:32:59	SELECT * FROM payment	16049 row(s) returned
3	16:33:00	SELECT * FROM rental	16044 row(s) returned

Figure 4-18 Source tables

Figure 4-19 shows the payment and rental table rows from the target database:

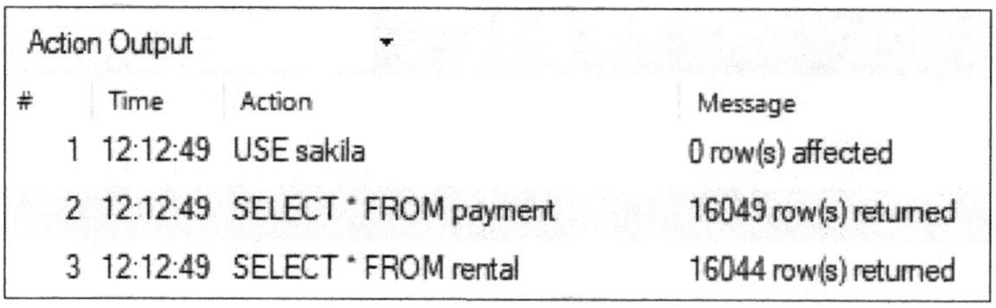

#	Time	Action	Message
1	12:12:49	USE sakila	0 row(s) affected
2	12:12:49	SELECT * FROM payment	16049 row(s) returned
3	12:12:49	SELECT * FROM rental	16044 row(s) returned

Figure 4-19 Target tables

4.2 MySQL to SQL Server Migration Example

We will use Microsoft SQL Server Migration Assistant for MySQL and ESF Database Migration Toolkit in this section.

4.2.1 Using Microsoft SQL Server Migration Assistant for MySQL Tool

Source Server and Database: MySQL 5.7 employees database
Target Server and Database: SQL Server 2016 employees databases
Migration Tool: Microsoft SQL Server Migration Assistant for MySQL

Steps to migrate from MySQL to SQL Server using SSMA for MySQL:

Step 1:

Open **SQL Server Migration Assistant for MySQL** then create a new project. Select SQL Server 2016 as target server. Click **OK**.

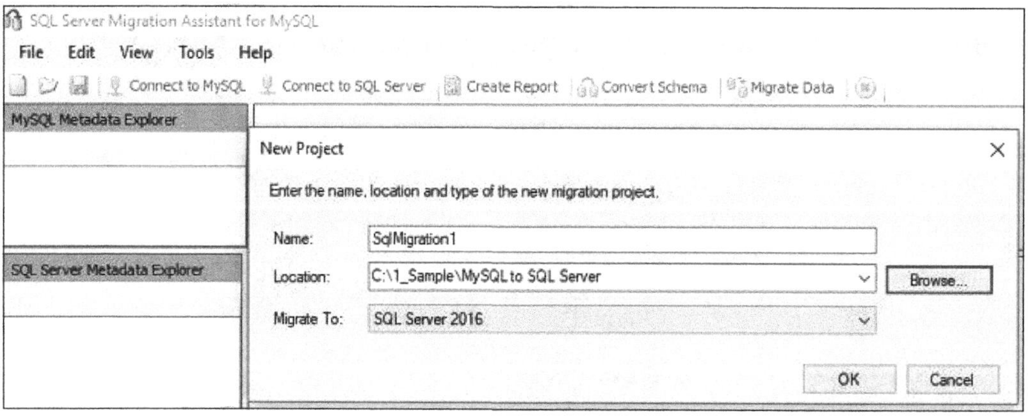

Figure 4-20 Creating new project in SQL Server Migration Assistant for MySQL

Step 2:

Click **Connect to MySQL** on top menu bar. Enter server name, user name and password. Click on **Connect button**.

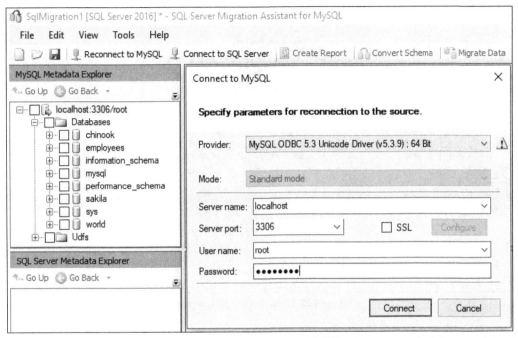

Figure 4-21 Connecting to MySQL

Step 3:

Click **Connect to SQL Server** on top menu bar. Enter database name **employees**.

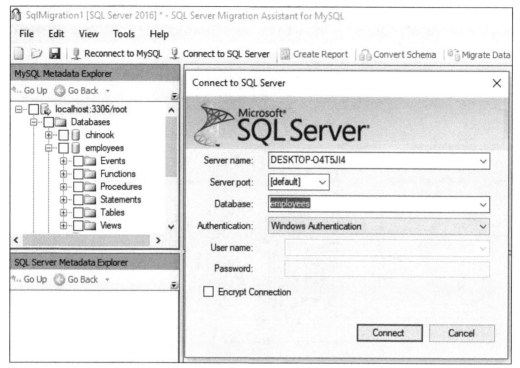

Figure 4-22 Connecting to SQL Server

A pop-up message shows up as we have not created table employees on SQL Server 2016: "Database 'employees' does not exist in the specified instance of SQL Server. Create database?"

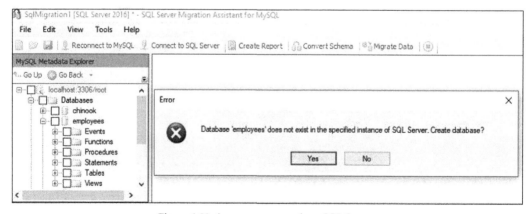

Figure 4-23 An error message from SQL Server

Step 4:

Click **Yes** to create the database:

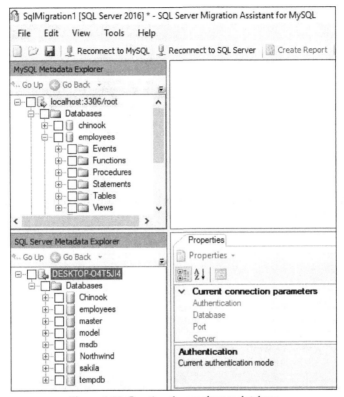

Figure 4-24 Creating the employees database

A second message appears: "SQL Server Agent is not running. You must start SQL Server Agent to use Server-side data migration engine."

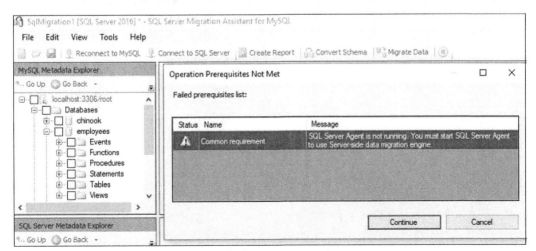

Figure 4-25 SQL Server agent warning message

Step 5:

Start the SQL Server Agent:

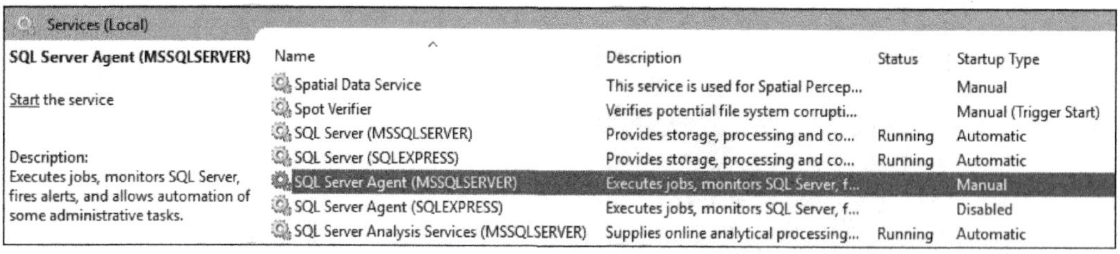

Services (Local)				
SQL Server Agent (MSSQLSERVER)	Name ^	Description	Status	Startup Type
	Spatial Data Service	This service is used for Spatial Percep...		Manual
Start the service	Spot Verifier	Verifies potential file system corrupti...		Manual (Trigger Start)
	SQL Server (MSSQLSERVER)	Provides storage, processing and co...	Running	Automatic
Description:	SQL Server (SQLEXPRESS)	Provides storage, processing and co...	Running	Automatic
Executes jobs, monitors SQL Server,	SQL Server Agent (MSSQLSERVER)	Executes jobs, monitors SQL Server, f...		Manual
fires alerts, and allows automation of some administrative tasks.	SQL Server Agent (SQLEXPRESS)	Executes jobs, monitors SQL Server, f...		Disabled
	SQL Server Analysis Services (MSSQLSERVER)	Supplies online analytical processing...	Running	Automatic

Figure 4-26 Starting the Agent

Step 6:

Click **Create Report** on the top menu. The report shows the migrated rows in salaries table are 2,597,046 (91.41% success rate).

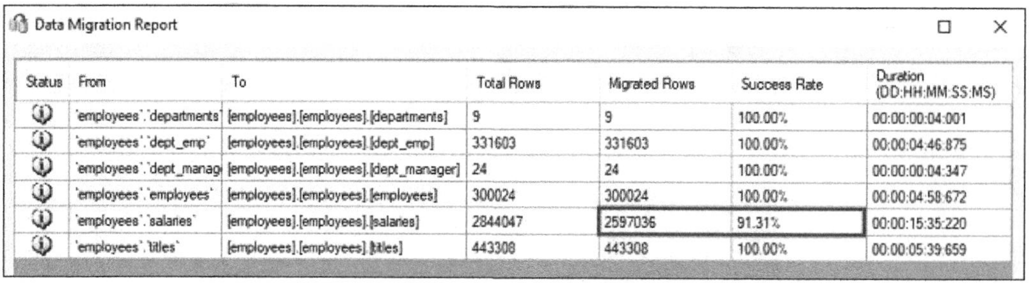

	Status	From	To	Total Rows	Migrated Rows	Success Rate	Duration (DD:HH:MM:SS:MS)
	⊕	`employees`.`departments`	[employees].[employees].[departments]	9	9	100.00%	00:00:00:04:001
	⊕	`employees`.`dept_emp`	[employees].[employees].[dept_emp]	331603	331603	100.00%	00:00:04:46:875
	⊕	`employees`.`dept_manag`	[employees].[employees].[dept_manager]	24	24	100.00%	00:00:00:04:347
	⊕	`employees`.`employees`	[employees].[employees].[employees]	300024	300024	100.00%	00:00:04:58:672
	⊕	`employees`.`salaries`	[employees].[employees].[salaries]	2844047	2597036	91.31%	00:00:15:35:220
	⊕	`employees`.`titles`	[employees].[employees].[titles]	443308	443308	100.00%	00:00:05:39:659

Figure 4-27 Data migration report

Step 7:

Click **Convert Schema** on the top menu. Once the **Conversion Statistics** show no error click **Migrate Data** on the top menu. Figure 4-29 shows the views are migrated automatically.

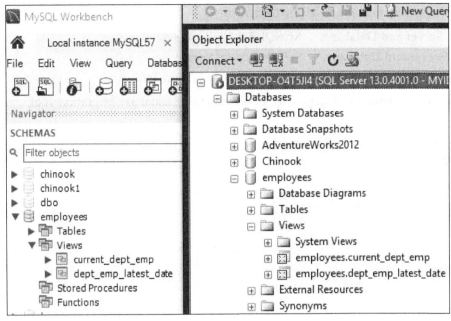

Figure 4-28 The comparison of views after migration

Figure 4-29 shows a comparison for employees and salaries table data types:

MySQL Tables	SQL Server Tables

Table: employees
 	
Columns:
emp_no int(11) PK
birth_date date
first_name varchar(14)
last_name varchar(16)
gender enum('M','F')
hire_date date
	⊟ 🗆 dbo.employees
	  ⊟ 🗀 Columns
	    🔑 emp_no (PK, int, not null)
	    🗉 birth_date (date, not null)
	    🗉 first_name (varchar(14), not null)
	    🗉 last_name (varchar(16), not null)
	    🗉 gender (varchar(1), not null)
	    🗉 hire_date (date, not null)

Table: salaries
 	
Columns:
emp_no int(11) PK
salary int(11)
from_date date PK
to_date date
	⊟ 🗆 dbo.salaries
	  ⊟ 🗀 Columns
	    🔑 emp_no (PK, FK, int, not null)
	    🗉 salary (int, not null)
	    🔑 from_date (PK, date, not null)
	    🗉 to_date (date, not null)

Figure 4-29 The comparison of data types after migration

4.2.2 Using ESF Database Migration Tool

Source Server and Database: MySQL 5.7 employees database

Target Server and Database: SQL Server 2016 employees1 databases

Migration Tool: ESF Database Migration Tool

Steps to migrate from MySQL to SQL Server using ESF Database Migration Tool:

Step 1:

Choose data Source **MySQL**. Enter the Server name and the password. Select database **employees**.

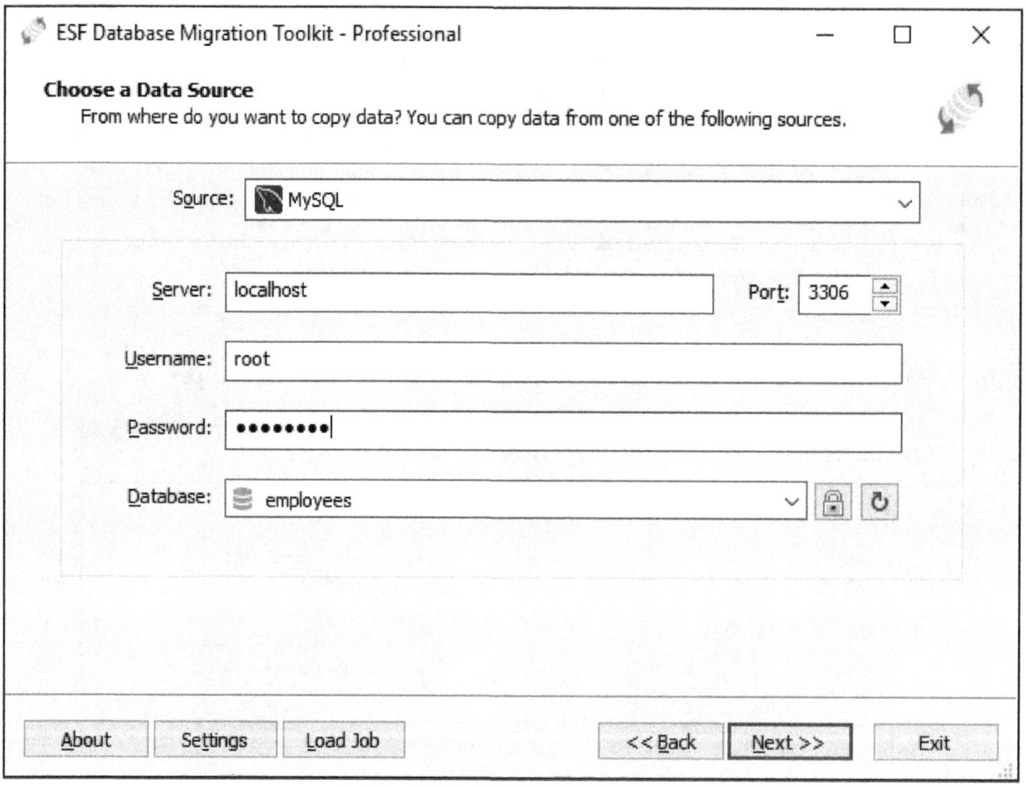

Figure 4-30 Choosing the source server

Step 2:

Create a database **employee1** on SQL Server 2016:

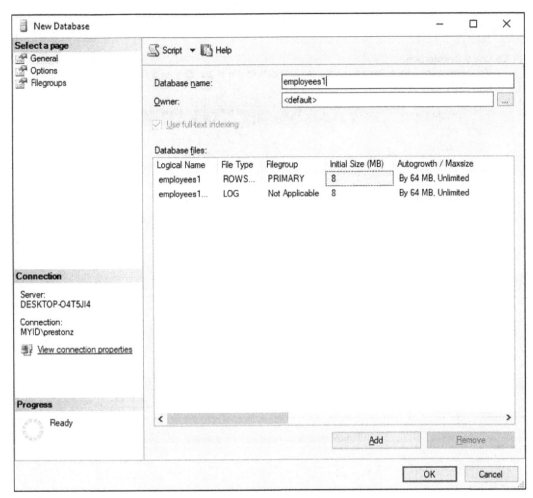

Figure 4-31 Creating a new database

Step 3:

Select Target server **SQL Server (Windows Authentication).** Enter server name and database name:

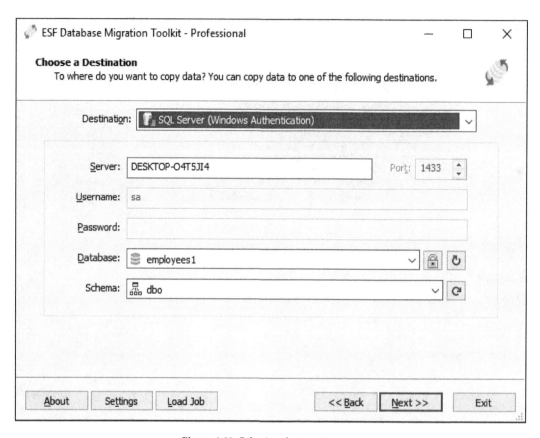

Figure 4-32 Selecting the target server

Step 4:

Click **Next**. All the source database tables and views appear. Check all the tables and views:

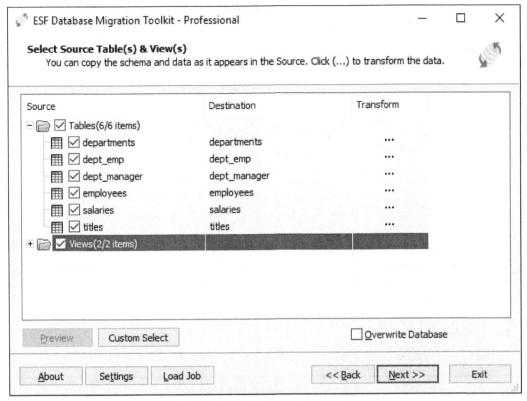

Figure 4-33 Selecting tables and views

Step 5:

Click on **Next button**.

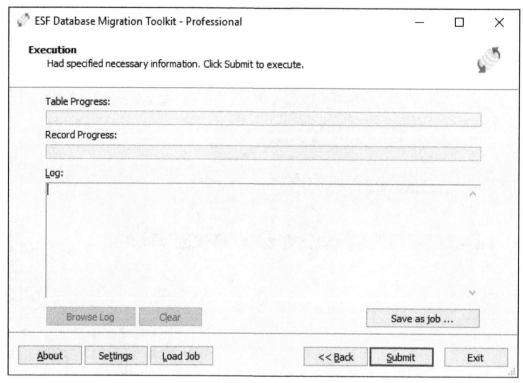

Figure 4-34 Executing the migration task

Step 6:

Click on **Submit** button. It took 1 hour 11 minutes for the employees1 database migration process.

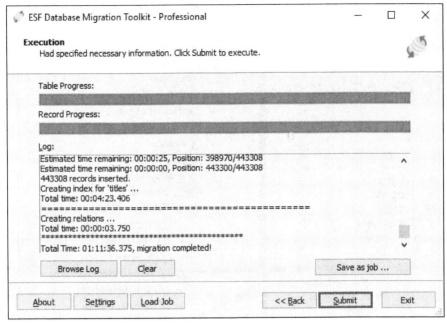

Figure 4-35 The migration progress and the log

Figure 4-36 shows employee1 tables records:

Table Name	# Records	Reserved (KB)
dbo.salaries	2,844,047	64,984
dbo.titles	443,308	14,808
dbo.dept_emp	331,603	12,592
dbo.employees	300,024	12,568
dbo.departments	9	144
dbo.dept_manager	24	144

Figure 4-36 The employee1 table records after migration

Figure 4-47 shows that two views in MySQL are migrated to SQL Server in tables:

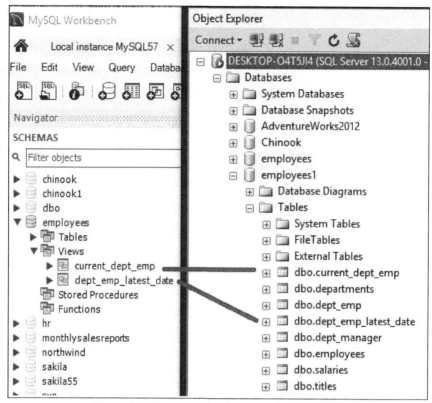

Figure 4-37 The comparison in views after migration

Figure 4-38 shows two table data types in source and target database after the migration. Note that MySQL enum type is changed to varchar(1) type in SQL Server.

MySQL Tables	SQL Server Tables

Table: employees
 	
Columns:
emp_no   int(11) PK
birth_date date
first_name varchar(14)
last_name varchar(16)
gender   enum('M','F')
hire_date date
	⊟ 📋 employees1
	  ⊞ 📁 Database Diagrams
	  ⊟ 📁 Tables
	    ⊞ 📁 System Tables
	    ⊞ 📁 FileTables
	    ⊞ 📁 External Tables
	    ⊟ 📄 dbo.departments
	      ⊟ 📁 Columns
	        🔑 dept_no (PK, char(4), not null)
	        📄 dept_name (varchar(40), not null)

Table: salaries
 	
Columns:
emp_no int(11) PK
salary int(11)
from_date date PK
to_date date
	⊟ 📄 dbo.salaries
	  ⊟ 📁 Columns
	    🔑 emp_no (PK, FK, int, not null)
	    📄 salary (int, not null)
	    🔑 from_date (PK, date, not null)
	    📄 to_date (date, not null)
	    ...

Figure 4-38 The comparison in data types

4.3 MySQL to Oracle Migration Example

Source Server and Database: MySQL 5.7 Employees database

Target Server and Database: Oracle 11g Employees database

Migration Tool: ESF Database Migration Toolkit

Steps to migrate from MySQL to Oracle using ESF Database Migration Toolkit:

Step 1:

Connect to Oracle database in Oracle SQL Developer. Right-click **Other Users** and create a user "EMPLOYEES":

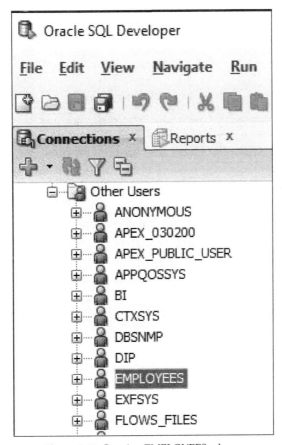

Figure 4-39 Creating EMPLOYEES schema

Step 2:

Open ESF Database Migration Tool. Choose "MySQL" as a source server. Enter username and password then select source database "employees":

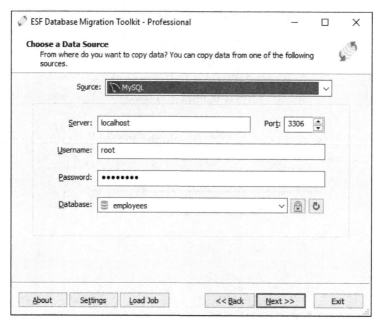

Figure 4-40 Choosing MySQL as the source server

Step 3:

Choose Oracle as destination server. Enter database connection information and database "EMPLOYEES":

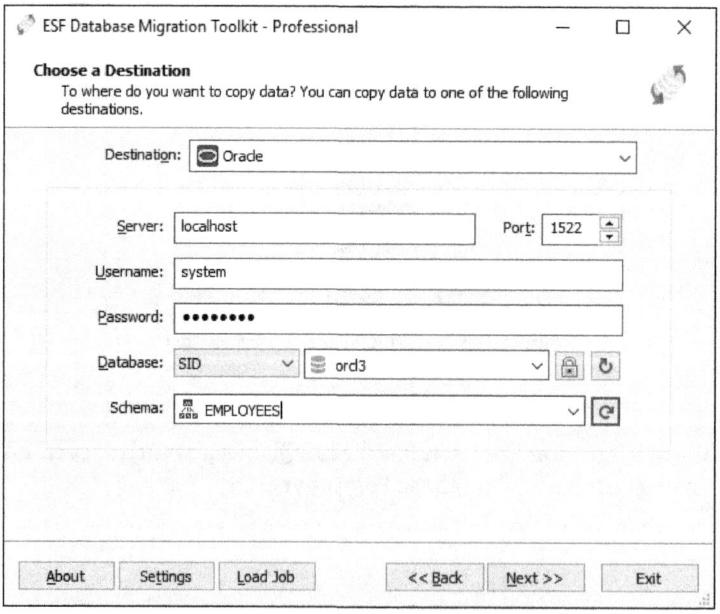

Figure 4-41 Choosing Oracle as the target server

Step 4:

Select source tables and views. We will not change field names and filter the data in this example.

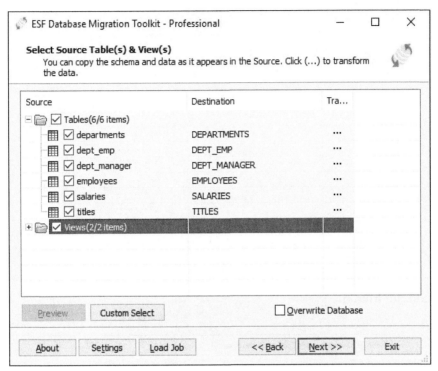

Figure 4-42 Selecting the source tables and views

Step 5:

Click on **Next button**.

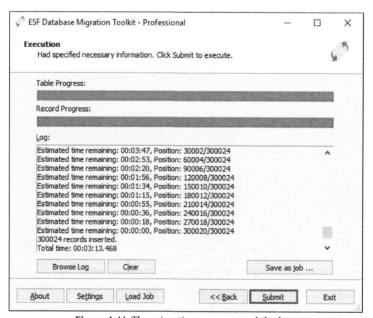

Figure 4-43 Executing the migration task

Step 6:

Click on **Submit button**.

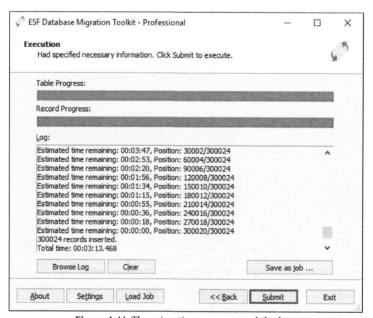

Figure 4-44 The migration progress and the log

Step 7:

Figure 4-45 shows that source **employees.salaries** table has 2,844,047 rows.

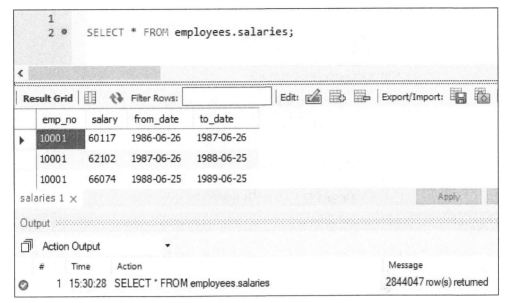

Figure 4-45 The source salaries table records

Step 8:

Enter the following SQL statements in the Query Builder window and execute the query:

SELECT table_name, num_rows counter
FROM dba_tables
WHERE owner = 'EMPLOYEES'
ORDER BY counter DESC

Figure 4-47 shows all the records from the target EMPLOYEES database:

	TABLE_NAME		COUNTER
1	SALARIES		2844047
2	TITLES		443308
3	DEPT_EMP		331603
4	EMPLOYEES		300024
5	DEPT_MANAGER		24
6	DEPARTMENTS		9

Figure 4-46 The target database table records

Step 9:

Right-click **dept_emp_latest_date -> Alter View** to get the following generated SQL code from the source MySQL database:

```
CREATE
        ALGORITHM = UNDEFINED
        DEFINER = `root`@`localhost`
        SQL SECURITY DEFINER
VIEW `dept_emp_latest_date` AS
        SELECT
                `dept_emp`.`emp_no` AS `emp_no`,
                MAX(`dept_emp`.`from_date`) AS `from_date`,
                MAX(`dept_emp`.`to_date`) AS `to_date`
        FROM
                `dept_emp`
        GROUP BY `dept_emp`.`emp_no`
```

Figure 4-47 Getting SQL code for the view

Step 10:

Copy the above SELECT statement from figure 4-47 to Notepad and replace all the " ` " character with a blank space:

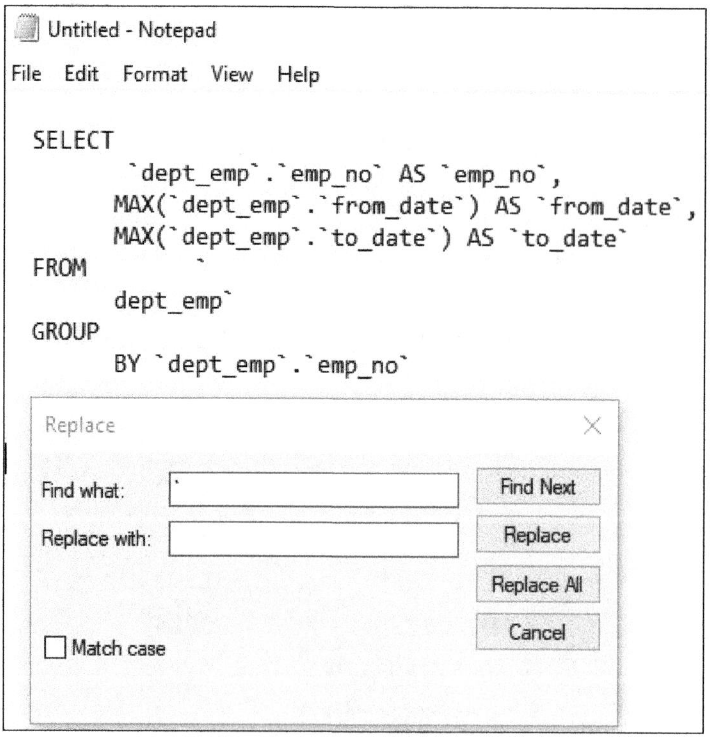

Figure 4-48 Editing the SQL code

Step 11:

Right-click **Views** and choose **New View** option in Oracle. Copy the modified SQL code from the Notepad and paste to the **SQL Query** part. Note that Employees schema name is added after FROM clause. Enter the name of the view:

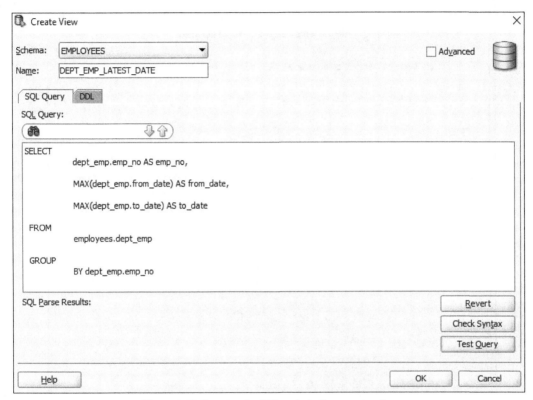

Figure 4-49 Copying the SQL code to the query section

Step 12:

Click **Test Query**. If no error, you will see the result: **Query executed successfully**.

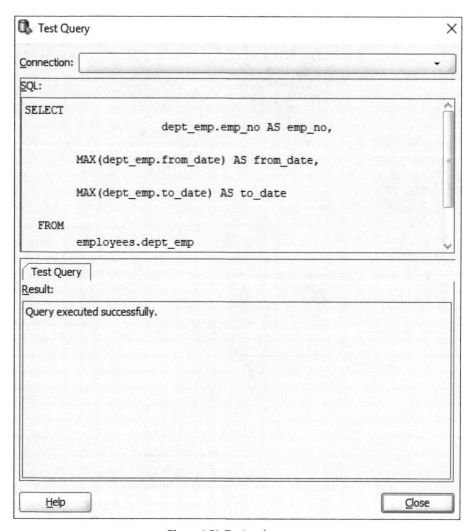

Figure 4-50 Testing the query

Step 13:

Close the **Test Query** window and click **Revert** button (see figure 4-50). The view **DEPT_EMP_LASTEST_DATE** is created in the target Oracle **EMPLOYEES** database.

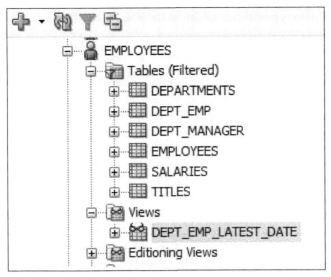

Figure 4-51 A view is created manually

Step 14:

Now we can drop the table **DEPT_EMP_LASTEST_DATE**:

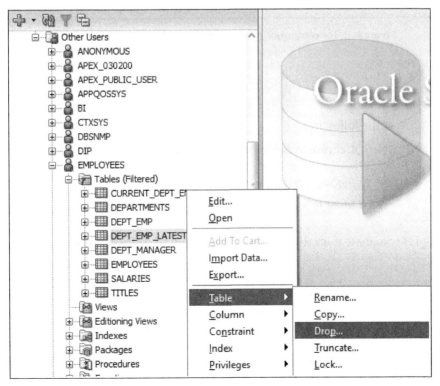

Figure 4-52 Dropping a table that is from a view

Figure 4-53 shows data types in two tables in the source and the target database:

MySQL Tables	Oracle Tables
Table: employees **Columns:** **emp_no** int(11) PK birth_date date first_name varchar(14) last_name varchar(16) gender enum('M','F') hire_date date	EMPLOYEES × Columns Data Constraints Grants Statistics Trigger ▼ Actions... COLUMN_NAME DATA_TYPE 1 EMP_NO NUMBER(38,0) 2 BIRTH_DATE DATE 3 FIRST_NAME VARCHAR2(14 BYTE) 4 LAST_NAME VARCHAR2(16 BYTE) 5 GENDER VARCHAR2(1 BYTE) 6 HIRE_DATE DATE
Table: salaries **Columns:** **emp_no** int(11) PK salary int(11) **from_date** date PK to_date date	SALARIES × Columns Data Constraints Grants Statistics ▼ Actions... COLUMN_NAME DATA_TYPE 1 EMP_NO NUMBER(38,0) 2 SALARY NUMBER(38,0) 3 FROM_DATE DATE 4 TO_DATE DATE

Figure 4-53 The comparison for the data types after the database migration

Summary

Chapter 4 covers the following:

- Using MySQL SQL commands for export and MySQL Import Wizard to migrate from MySQL to MySQL.
- Using MySQL Workbench Database Schema Transfer Wizard.
- Using MSMA for MySQL and EST database migration tool to migrate from MySQL to SQL Server
- Using ESF database migration toolkit to migrate from MySQL to Oracle

Chapter 5

More About MySQL Workbench

MySQL Workbench is a powerful tool for developers and DBAs to design MySQL databases, write SQL statements, manage MySQL databases and migrate databases. We have seen the migration wizard in MySQL Workbench in the last chapter. There are other features in MySQL Workbench, for example, Data Modeling and Database Management.

5.1 MySQL Database Modeling

5.1.1 Visual Database Design

Steps for visual database design:

Step 1:
Click on the "+" sign next to the **Models** to create a new model:

Figure 5-1 New model in MySQL workbench

Step 2:

Right-click the default **mydb** schema and select **Edit Schema**:

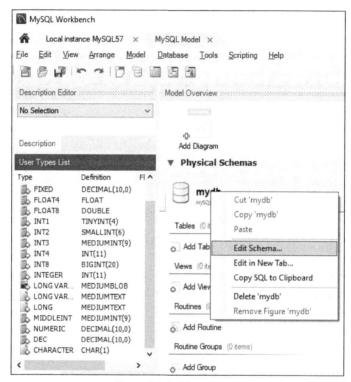

Figure 5-2 Editing schema

Step 3:

Enter a new schema name **course** and add a table **courses** with three fields:

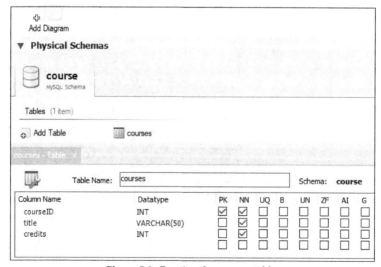

Figure 5-3 Creating the courses table

Step 4:

Double click **Add Table** icon to create course_instructor table with two fields:

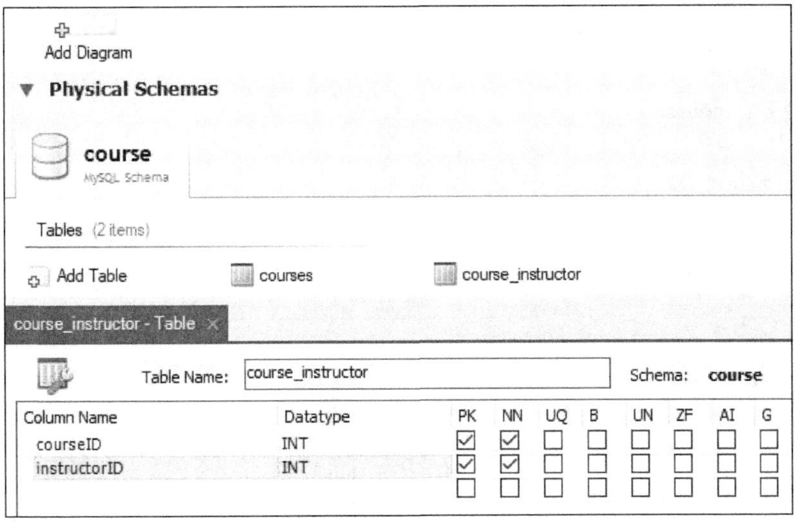

Figure 5-4 Creating the course_instructor table

Step 5:

Double click **Add Table** icon to create **instructor** table with four fields:

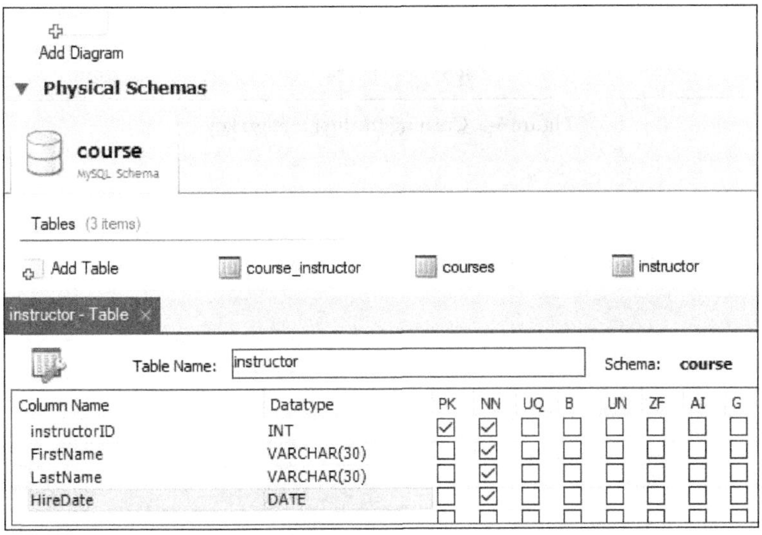

Figure 5-5 Creating the instructor table

Step 6:

Click **course_instructor** table then click **Foreign Keys** tab at the button. Create the first foreign key:

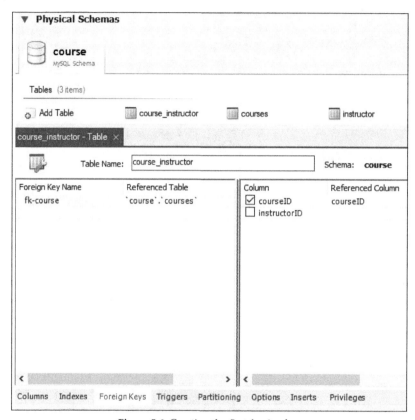

Figure 5-6 Creating the first foreign key

Step 7:

Add the second foreign key **fk-instructor**:

Figure 5-7 Creating the second foreign key

Step 8:

Double click **Add Diagram** icon above the course schema then drag the three tables on the left pane to the new diagram. The ER diagram shows the correct relationships between the three tables:

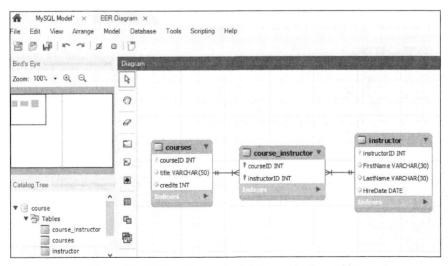

Figure 5-8 Relationships are built between the three tables

5.1.2 *Reverse Engineering*

Sometimes we need to create a diagram from an existing database. This process is called Reverse Engineering.

Steps to do the Reverse Engineering:

Step 1:

Click **Database** -> **Reverse Engineer** on the top menu:

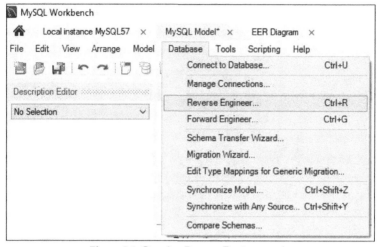

Figure 5-9 Opening Reverse Engineer tool

Step 2:

MySQL identifies hostname, port number and username. Click on the **Next** button:

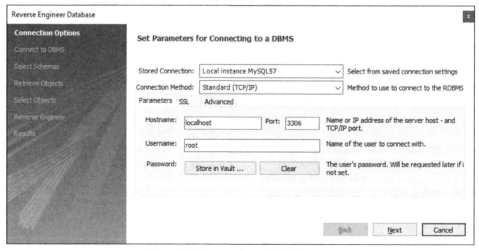

Figure 5-10 Connecting to MySQL server

Step 3:

MySQL Workbench connects to the database instance and gets a list of databases. Click **Next:**

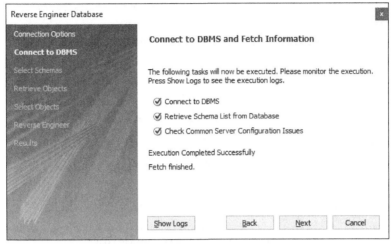

Figure 5-11 Retrieving schema list from the server

Step 4:

Choose the database that you want to reverse engineer, then click **Next:**

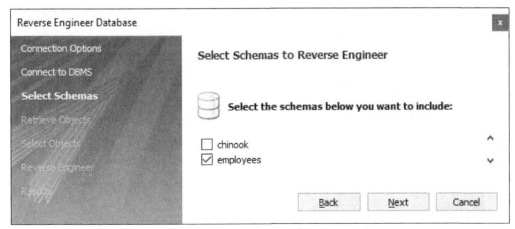

Figure 5-12 Selecting the database for reverse engineer

Step 5:

Workbench will retrieve objects from the selected schema. Click on the **Next** button:

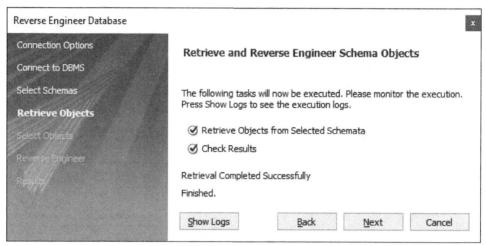

Figure 5-13 Retrieving objects from the selected database

Step 6:

You can select tables that you need to reverse engineer, then click on the **Execute>** button:

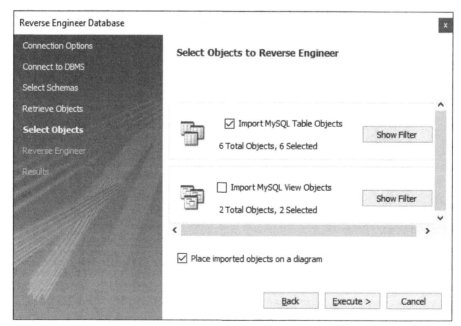

Figure 5-14 Selecting tables or views for reverse engineer

Step 7:

Workbench will show reverse engineering progress. Click on the **Next** button:

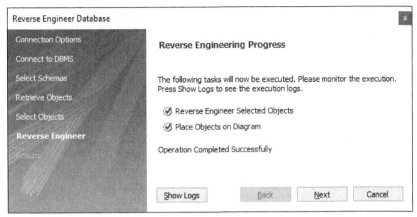

Figure 5-15 Reverse engineer status

Step 8:

You will see a relational diagram on the screen. Below is the employees database diagram:

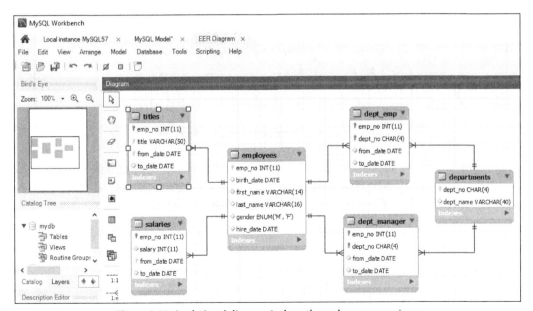

Figure 5-16 A relational diagram is done through reverse engineer

5.1.3 A Visual Query Builder for MySQL and Other Databases

Since MySQL Workbench does not has a visual query builder available, I would recommend using **FlySpeed SQL Query** for creating MySQL queries. **FlySpeed SQL Query** can be used for all the major database systems. It's a free tool if you do not print and export data.

Steps for query design in **FlySpeed SQL Query**:

Step 1:

Open **FlySpeed SQL Query** and click **Connection** on the top menu:

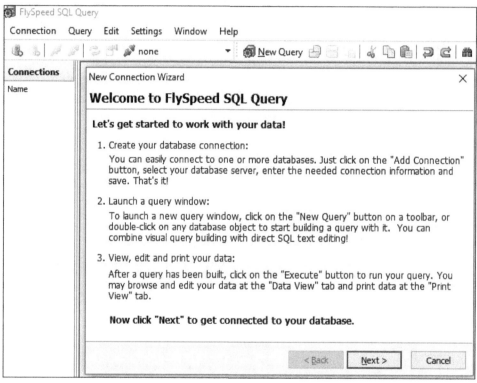

Figure 5-17 FlySpeed SQL Query Welcome screen

Step 2:

Select **MySQL** as database server:

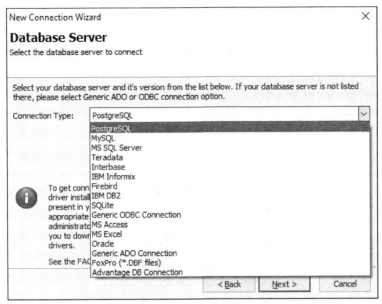

Figure 5-18 Selecting a database server

Step 3:

Go to employees schema. Drag employees, dep_emp and departments tables from the left pane to the main screen. You can enter query filters if needed:

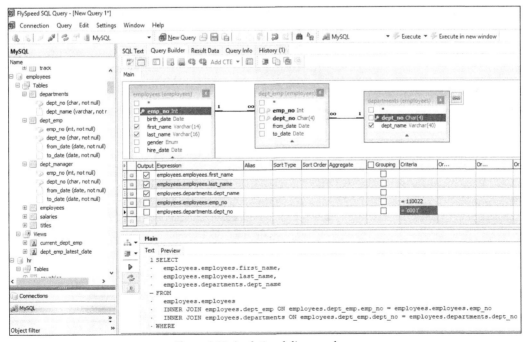

Figure 5-19 A relational diagram shows

Below is the generated SQL code at the bottom of the main screen:

```
SELECT
    employees.employees.first_name,
    employees.employees.last_name,
    employees.departments.dept_name
FROM
    employees.employees
INNER JOIN employees.dept_emp ON employees.dept_emp.emp_no =
employees.employees.emp_no
    INNER JOIN employees.departments ON employees.dept_emp.dept_no =
employees.departments.dept_no
    WHERE
        employees.employees.emp_no = 110022 AND
        employees.departments.dept_no = 'd001'
```

Figure 5-20 Generated SQL code

Step 4:

You can see the query result by clicking the **Execute** button at the top screen:

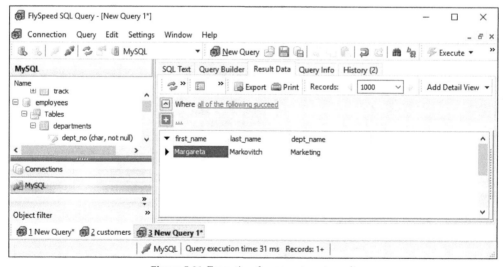

Figure 5-21 Executing the query to get result

5.2 MySQL Server Management

MySQL Workbench has a tree-based navigation that allows administrators to view server connections. There are four sections in the management pane: MANAGEMENT, INSTANCE, PERFORMANCE and MySQL ENTERPRISE.

5.2.1 Server Status

Click **Server Status** in the Management section to display the Server Status screen:

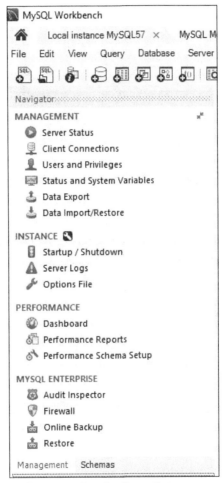

Figure 5-22 Opening Server Status page

The server status screen shows the instance name, version, server features and directories. It also has information for server traffic and load, etc.

Figure 5-23 Detail of Server Status page

5.2.2 *Server Logs*

Server Logs are located at INSTANCE section. It has Error Log File tag and Slow Log File tag. Figure 5-16 displays an example of the Show Log File:

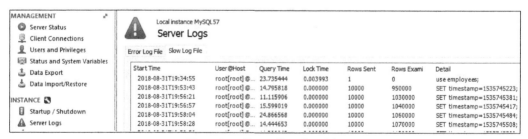

Figure 5-24 Server Show Log File

5.2.3 *Server Performance Dashboard*

To open the Server Performance Dashboard, click **Dashboard** from the **PERFORMANCE** section in the left pane. The dashboard displays statistics for network traffic and MySQL server performance status:

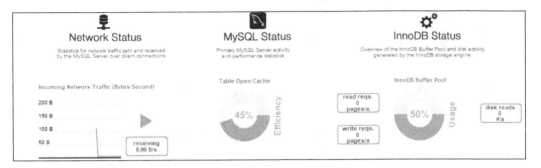

Figure 5-25 Server Performance Dashboard page

5.3 Online Backup

To perform MySQL Enterprise Backup we need to create a user MySQLBackup with BackupAdmin role:

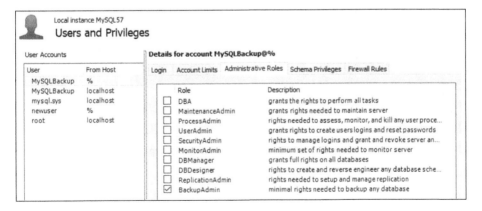

Figure 5-26 Creating a user MySQLBackup

To do a backup click **Online Backup** from **MySQL ENTERPRISE** section in the left pane. When doing the backup first time Workbench will prompt a message asking for fixing MySQL Enterprise Backup grants. Click the Fix Grants for MEB button:

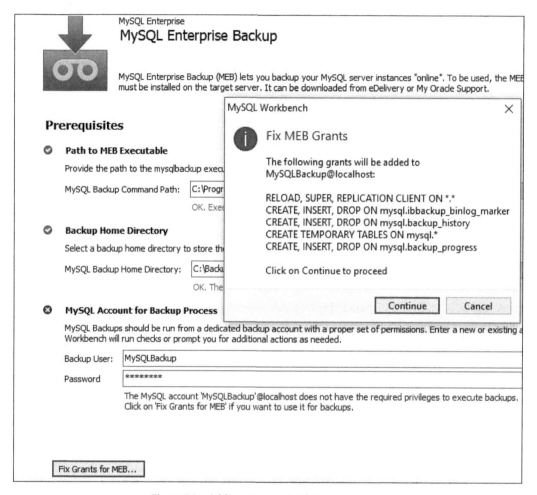

Figure 5-27 Adding grants to MySQLBackup user

Click the **Continue** button to schedule a full backup:

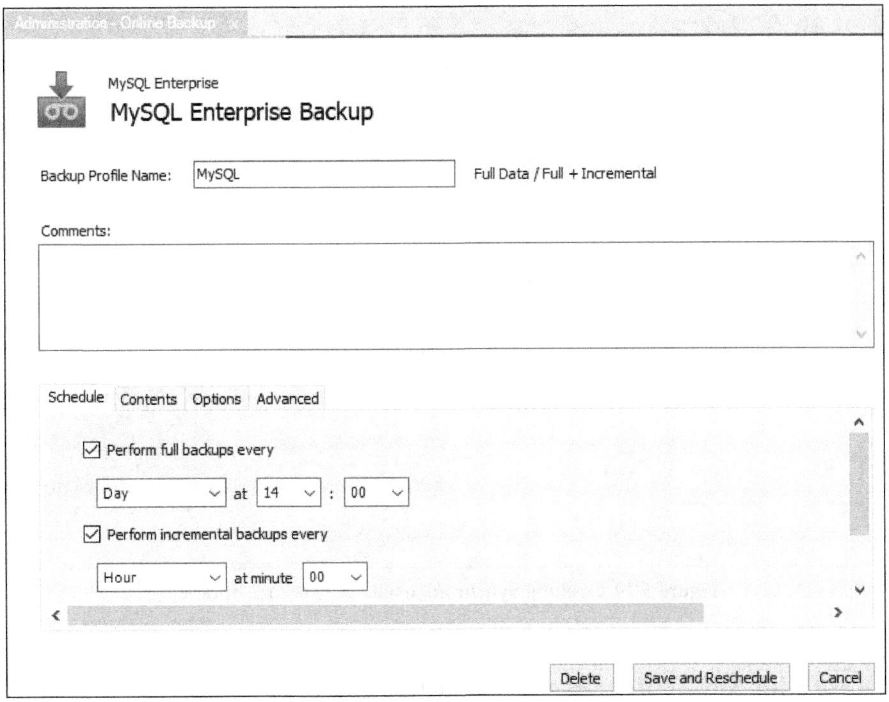

Figure 5-28 Full backup schedule

5.4 MySQL Synchronization

MySQL Workbench's Synchronization tool can compare a MySQL server with a target MySQL server and perform synchronization between the two servers. It's a good schema comparison tool for a production server with a development server.

Steps to do synchronization between two servers:

Step 1:

Open a model then select **Database -> Synchronize with Any Source** option:

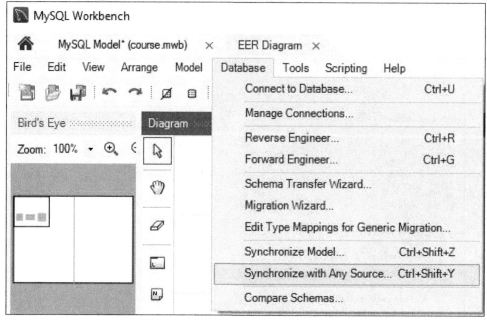

Figure 5-29 Opening Synchronize with Any Source option

Synchronize with Any Source has an introduction page:

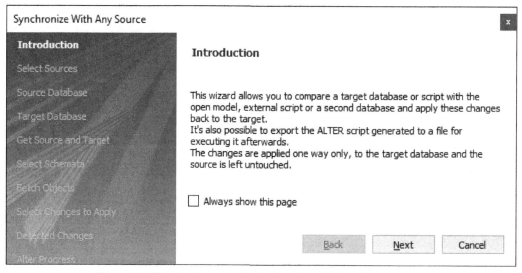

Figure 5-30 Introduction page for Synchronize with Any Source option

Step 2:

Verify the source server username and password:

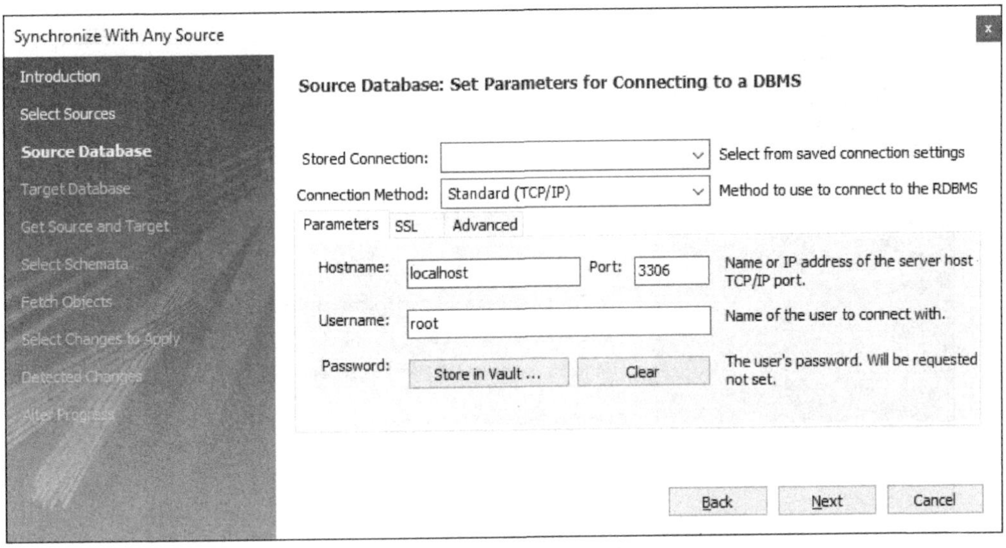

Figure 5-31 Connecting to the source server

Step 3:

Verify the target server username and password:

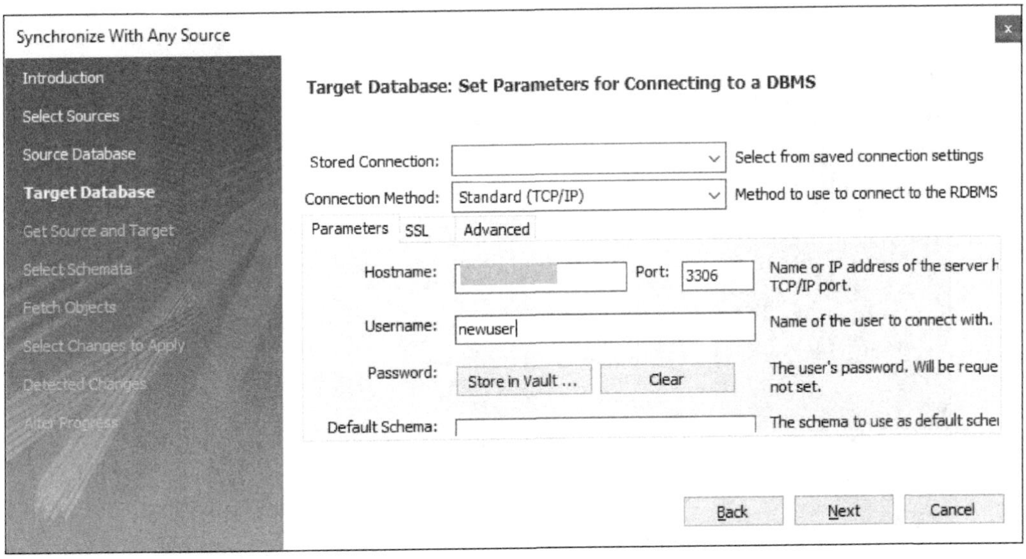

Figure 5-32 Connecting to the target server

Step 4:

The synchronize tool will connect the two servers and get all the databases from the two servers:

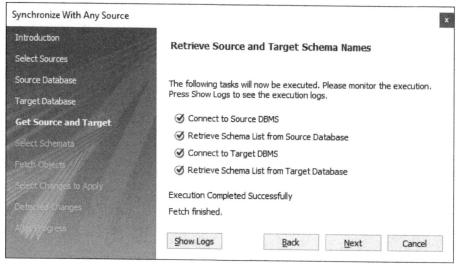

Figure 5-33 Retrieving schema names from the source and the target server

Step 5:

The tools can check if the schemas match or not. Click **Next** to select a schema to be synchronized:

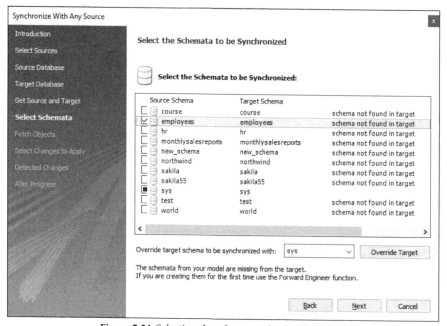

Figure 5-34 Selecting the schemas to be Synchronized

Step 6:

The synchronized tool found the difference between the two servers for the schema you selected:

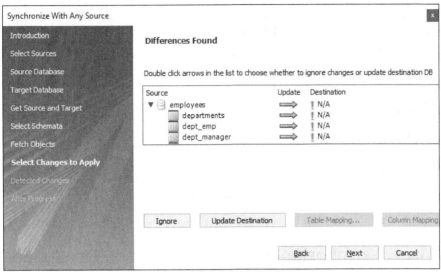

Figure 5-35 You can choose upgrading the target database or not

Step 7:

Click **Next** to finish the synchronize process. Below are the databases that the synchronize tool created on the target server:

Figure 5-36 The synchronize tool created database on the target server

5.5 Unique MySQL INSERT IGNORE Statement

MySQL Insert Ignore statement is very useful when dealing with an automated incoming file. For example, we receive a file every night from another company. We need to insert the records from the file to our location before 8am. However, the incoming file has some duplicated records. If we use regular INSERT statement, we will get a duplicate entry error message and the insert process will be stopped:

> *INSERT INTO employees.departments (dept_no, dept_name)*
> *VALUES ('d005', 'Development');*

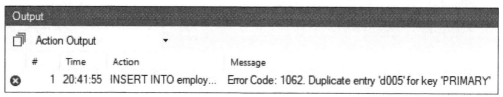

Figure 5-37 Using INSERT INTO statement

When we use INSERT IGNORE statement, we only get a warning message and the insert process will continue:

> *INSERT **IGNORE** INTO employees.departments (dept_no, dept_name)*
> *VALUES ('d005', 'Development');*

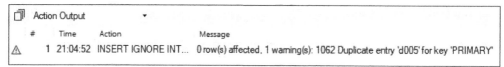

Figure 5-38 Using INSERT IGNORE INTO statement

Summary

Chapter 5 covers the following:

- Visual database design in MySQL Workbench
- Reverse engineer in MySQL Workbench
- A visual query builder for MySQL and other databases (FlySpeed SQL Query)
- MySQL Server Status
- MySQL Server Logs
- MySQL Server Performance Dashboard
- MySQL Online Backup
- Synchronization in MySQL Workbench

Chapter 6

Oracle Database Migration

The main topics in this chapter are illustrated in Figure 6-1: Oracle database migration to MySQL, Oracle database migration to SQL Server and Oracle database migration to Oracle.

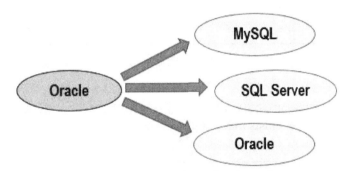

Figure 6-1 Oracle database migration paths

6.1 Oracle to MySQL Migration Example

Prerequisites:	Windows XP/Vista/7/8/8.1/10
	Oracle 9i/10g/11g/12c
	MySQL 3.23 and above
Source Server and Database:	Oracle 11 g HR database
Target Server and Database:	MySQL 5.7 HR database
Migration Tool:	ESF Database Migration Toolkit

Steps to migrate from Oracle to MySQL:

Step 1:

Open ESF Database Migration Toolkit and connect to the source Oracle database. Enter Oracle SID **orcl3** and schema name **HR:**

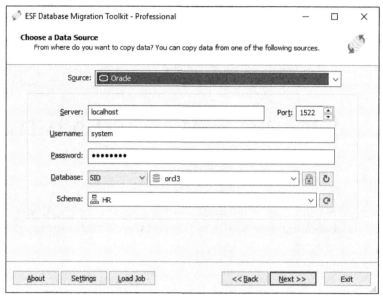

Figure 6-2 Connecting the source Oracle server

Step 2:

Choose **MySQL** as a target server. Enter database name **hr**. The toolkit will create the database if **hr** database not exists in MySQL server.

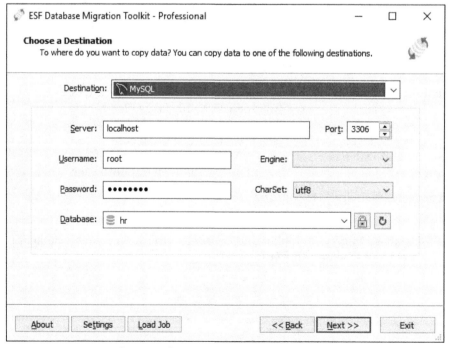

Figure 6-3 Connecting the target MySQL server

Step 3:

Select source tables that need to be migrated.

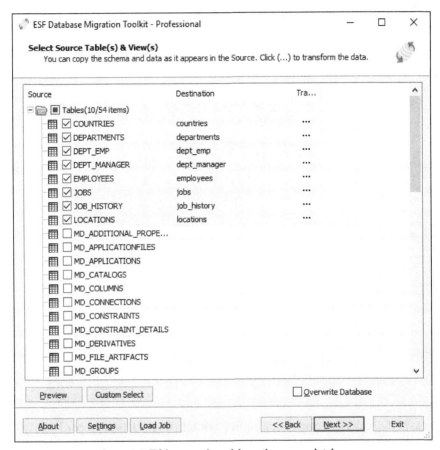

Figure 6-4 Tables are selected from the source database

Step 4:

Click **Next** then click **Submit**.

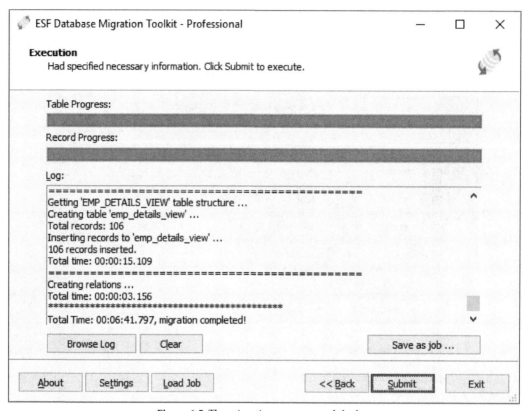

Figure 6-5 The migration progress and the log

Step 5:

Open Oracle SQL Developer and check source HR.DEPT_EMP table records: 331,603 rows:

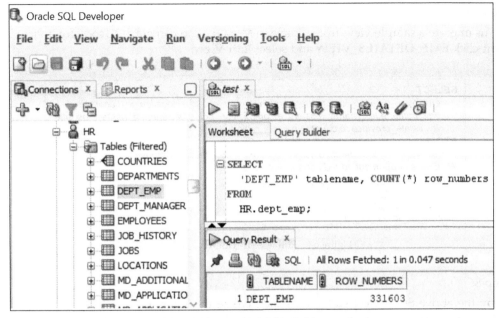

Figure 6-6 The source database table records

Step 6:

Open MySQL Workbench and display target **hr.dept_emp** table records: 331,603.

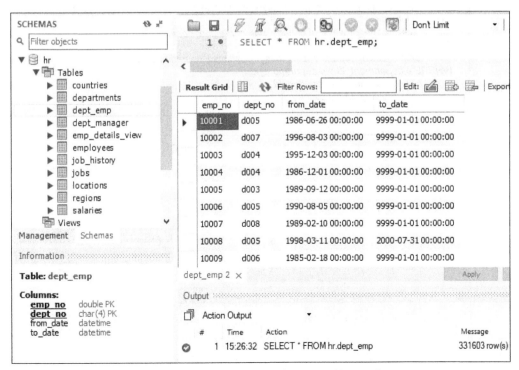

Figure 6-7 The target database dep_emp table records

Step 7:

Let us migrate a sample view from Oracle to MySQL. You can get SQL code for a view by right-click **EMP_DETAILS_VIEW** and select **Edit View**:

```
SELECT
    e.employee_id, e.job_id, e.manager_id, e.department_id, d.location_id, l.country_id,
    e.first_name, e.last_name, e.salary, e.commission_pct, d.department_name, j.job_title,
    l.city, l.state_province, c.country_name, r.region_name
FROM
    employees e, departments d, jobs j, locations l, countries c, regions r
WHERE e.department_id = d.department_id
    AND d.location_id = l.location_id
    AND l.country_id = c.country_id
    AND c.region_id = r.region_id
    AND j.job_id = e.job_id
```

Figure 6-8 Editing EMP_DETAILS_VIEW

Step 8:

Copy the above SQL select statements then create the view in MySQL by entering the following commands in the Query window. Run the query to create the view:

```
CREATE VIEW HR.EMP_DETAILS_VIEW AS
    SELECT e.employee_id, e.job_id, e.manager_id, e.department_id, d.location_id,
            l.country_id,
            e.first_name, e.last_name, e.salary, e.commission_pct, d.department_name,
            j.job_title,
            l.city, l.state_province, c.country_name, r.region_name
    FROM
            employees e, departments d, jobs j, locations l, countries c, regions r
    WHERE e.department_id = d.department_id
            AND d.location_id = l.location_id
            AND l.country_id = c.country_id
            AND c.region_id = r.region_id
            AND j.job_id = e.job_id
```

Figure 6-9 Creating a new view in MySQL

6.2 Oracle to SQL Server Migration Example

Prerequisites:	Windows XP/Vista/7/8/8.1/10
	Windows Server 2008 R2/2012/2012 R2/2016
	Oracle 9i and above
Source Server and Database:	Oracle 11g HR database
Target Server and Database:	SQL Server 2016 HR database
Migration Tool:	Microsoft SQL Server Migration Assistant for Oracle

Steps to migrate from Oracle to SQL Server:

Step 1:

Let us download the SQL Server Migration Assistant for Oracle at:
https://www.microsoft.com/en-us/download/details.aspx?id=54258

Figure 6-10 SQL Server Migration Assistant for Oracle download page

Step 2:

After the installation start the SQL Server Migration Assistant for Oracle by double click the icon:

Figure 6-11 SSMA for Oracle

Step 3:

You need to Install SQL Server Migration Assistant for Oracle Extension Pack on the target SQL server.

Step 4:

Click **File -> New Project** to create a new project. Choose a location for the project file and set target SQL Server:

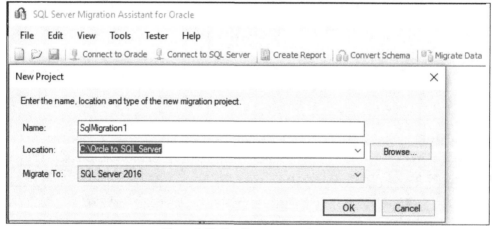

Figure 6-12 Creating a new project

Step 5:

Select **Default Project Settings** from Tools on the top menu. Click **Loading System Objects** then check **HR** database. Click **OK**.

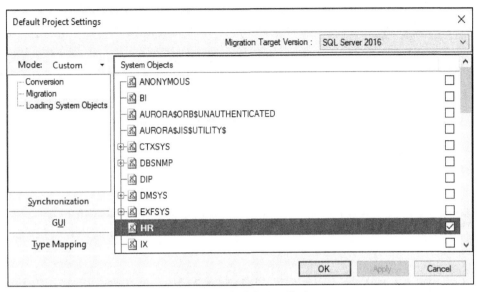

Figure 6-13 Default Project Settings

Step 6:

Click the **Connect to Oracle** from the toolbar. Enter source Oracle server name, port, Oracle SID, etc. Click **Connect** button.

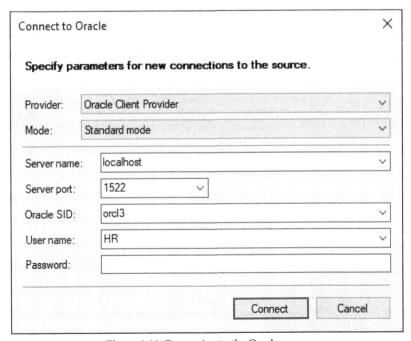

Figure 6-14 Connecting to the Oracle server

It will take a while to load objects:

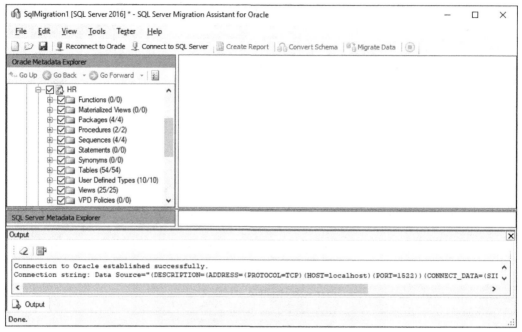

Figure 6-15 Oracle HR objects are loaded

Step 7:

Right-click **HR** schema and choose **Create Report**:

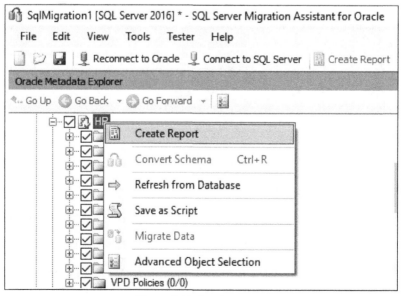

Figure 6-16 Creating report

Step 8:

Make sure that no error messages in the conversion statistics. Click **Connect to SQL Server** from the top menu. Enter the server, port number, database name. Click **Connect** button.

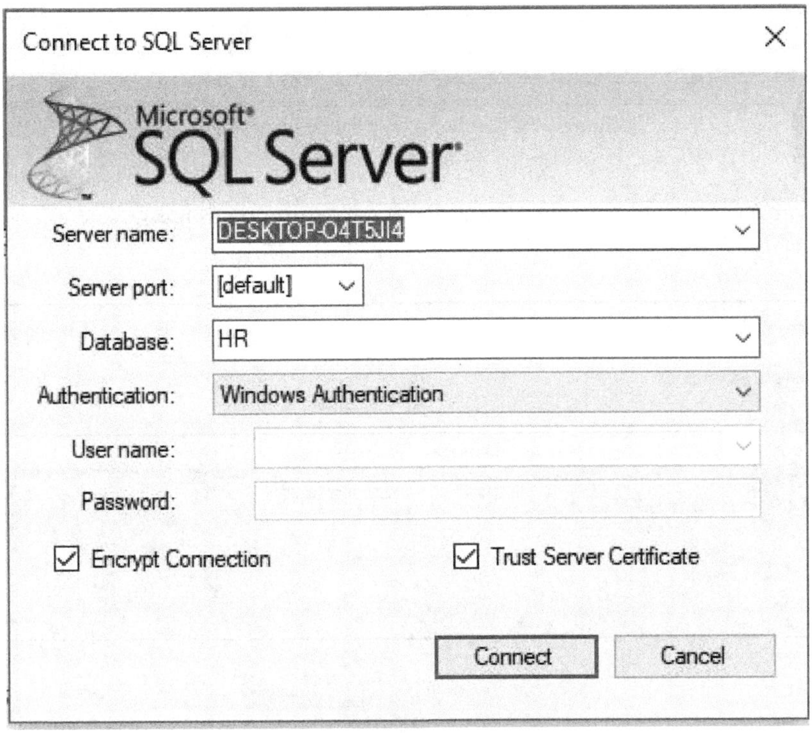

Figure 6-17 Connecting to SQL Server

If the database HR does not exist, SQL Server Migration Assistant for Oracle will ask for creating a new database.

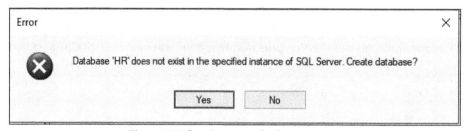

Figure 6-18 Creating a new database message

If SQL Server Agent is not running you will get a warning message:

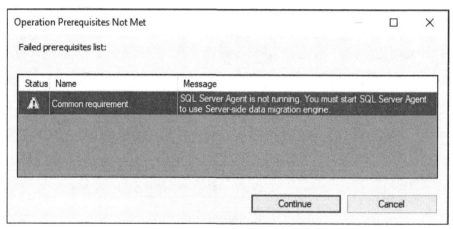

Figure 6-19 SQL Server Agent message

Step 9:

Go to SQL Server Management Studio. Right-click the **SQL Server Agent** and choose **Start**:

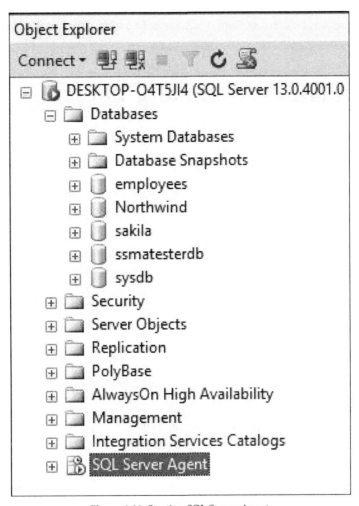

Figure 6-20 Starting SQL Server Agent

Step 10:

Right-click the **target HR** database and select **Synchronize with Database:**

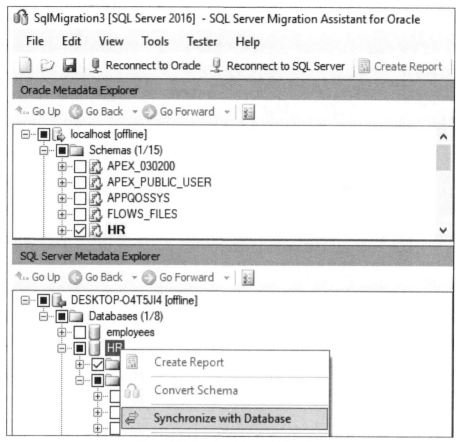

Figure 6-21 Synchronizing with Database

Step 11:

Right-click the **source HR** database and select **Migrate Data**:

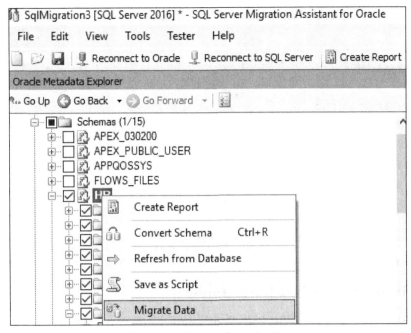

Figure 6-22 Starting data migration

When the migration is done a report appears. If the success rates are not 100% then you need to check the migration steps.

Status	From	To	Total Rows	Migrated Rows	Success Rate	Duration (DD:HH:MM:SS:MS)
ⓘ	"HR"."JOB_HISTORY"	[HR].[dbo].[JOB_HISTOR	10	10	100.00%	00:00:00:25:708
ⓘ	"HR"."EMPLOYEES"	[HR].[dbo].[EMPLOYEES]	107	107	100.00%	00:00:00:27:773
ⓘ	"HR"."LOCATIONS"	[HR].[dbo].[LOCATIONS]	23	23	100.00%	00:00:00:18:671
ⓘ	"HR"."REGIONS"	[HR].[dbo].[REGIONS]	4	4	100.00%	00:00:00:17:343
ⓘ	"HR"."DEPARTMENTS"	[HR].[dbo].[DEPARTMEN	27	27	100.00%	00:00:00:26:775
ⓘ	"HR"."COUNTRIES"	[HR].[dbo].[COUNTRIES]	25	25	100.00%	00:00:00:26:552
ⓘ	"HR"."JOBS"	[HR].[dbo].[JOBS]	19	19	100.00%	00:00:00:19:563

Figure 6-23 Data migration report

Figure 6-25 shows the target HR database table records:

Table Name	# Records
dbo.EMPLOYEES	107
dbo.JOB_HISTORY	10
dbo.LOCATIONS	23
dbo.DEPARTMENTS	27
dbo.COUNTRIES	25
dbo.JOBS	19
dbo.REGIONS	4

Figure 6-24 The employees table on the target SQL Server database

6.3 Oracle to Oracle Migration Example

6.3.1 Using Oracle Data Dump Export and Import in SQL Developer

After upgrading Oracle database, you can use Oracle SQL Developer Data Pump tool for database migration.

Steps for Data Dump Export (expdp) in SQL Developer:

Step 1:

Create a directory for export and grant privileges to a schema that needs to be exported:

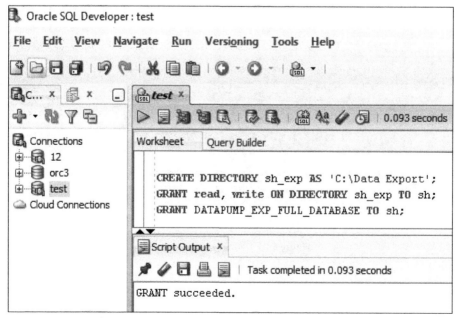

Figure 6-25 Creating a directory and grant privileges to SH

Step 2:

Click **View** -> **DBA** then right-click **Data Pump** -> **Data Pump Export Wizard**:

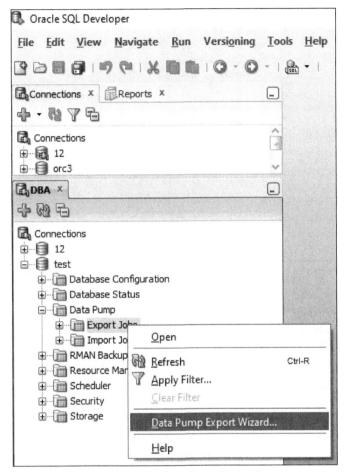

Figure 6-26 Opening Data Pump Export Wizard

Step 3:

We want to export **SH** schema, so we select **Schemas** type:

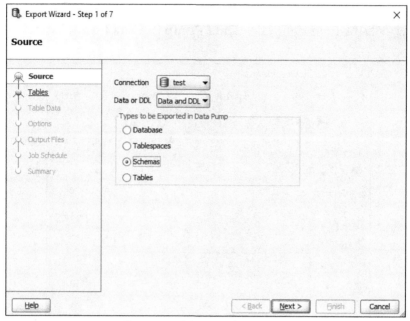

Figure 6-27 Selecting Schemas type

Step 4:

We select the **SH** schema in the left pane, then click the ">" button to move the schema to the right pane. Click on the **Next** button.

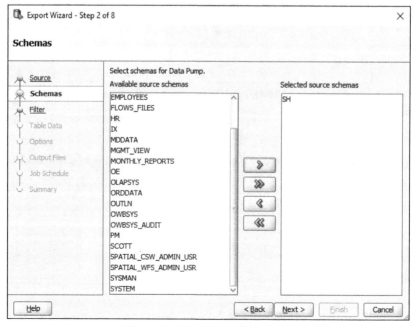

Figure 6-28 Moving the SH schema

Step 5:

You can apply include exclude filter to the schema. Click on the **Next** button:

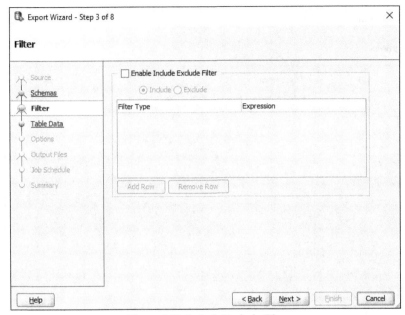

Figure 6-29 Include Exclude Filter

Step 6:

You can apply WHERE clause at this step. Click on the **Next** button:

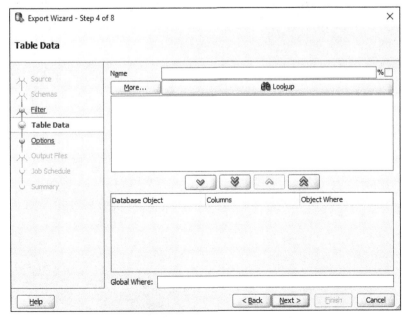

Figure 6-30 WHERE clause (optional)

Step 7:

Select the export directory in the Log File drop-down list. Click on the **Next** button:

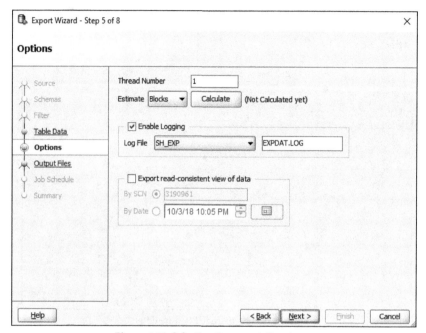

Figure 6-31 Selecting the export directory

Step 8:

You can double-click the **File Names** to change the dump file name. Click on the **Next** button:

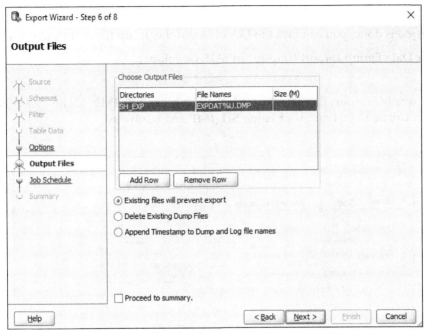

Figure 6-32 Changing the dump file name if needed

Step 9:

Set up a time to start the job. Here we select to start the job **Immediately**. Click on the **Next** button:

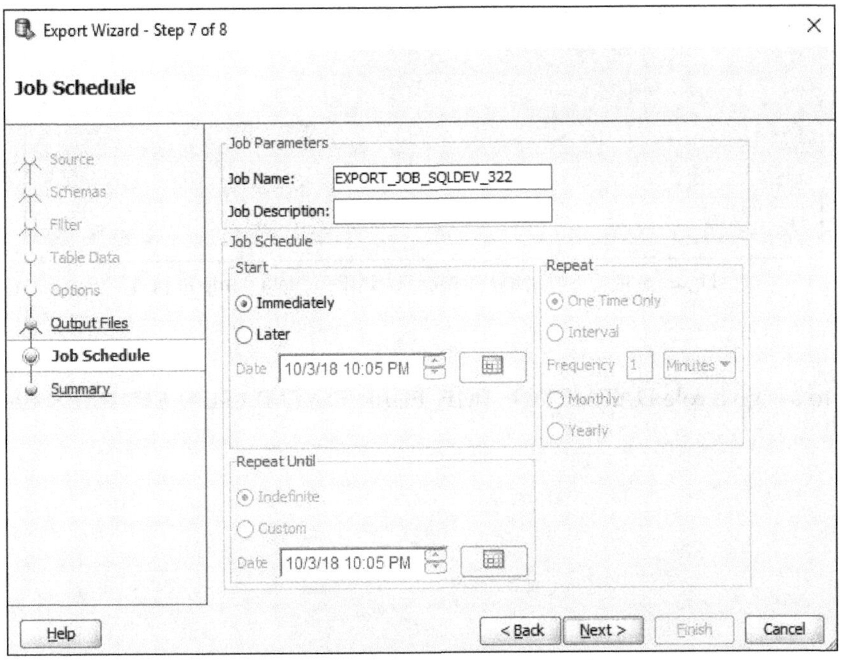

Figure 6-33 Scheduling the output job

Step 10:

Once the job is done, you can find the LOG file and the dump file in the export directory.

Steps for Data Dump Import (impdp) in SQL Developer:

Step 1:

Now we want to import the **SH** schema into a new user **SH_IMP**. Right-click other users and select Create User. Enter user name **SH_IMP** and password:

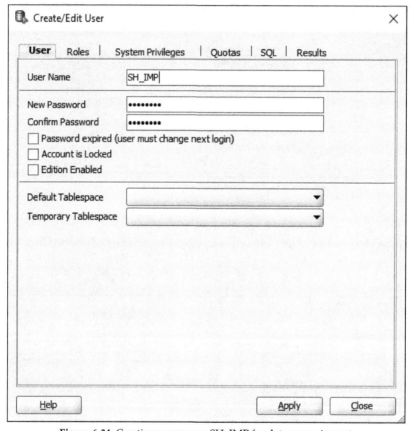

Figure 6-34 Creating a new user SH_IMP for data pump import

Step 2:

We need to assign a role **DATAPUMP_IMP_FULL_DATABASE** to **SH_IMP** schema:

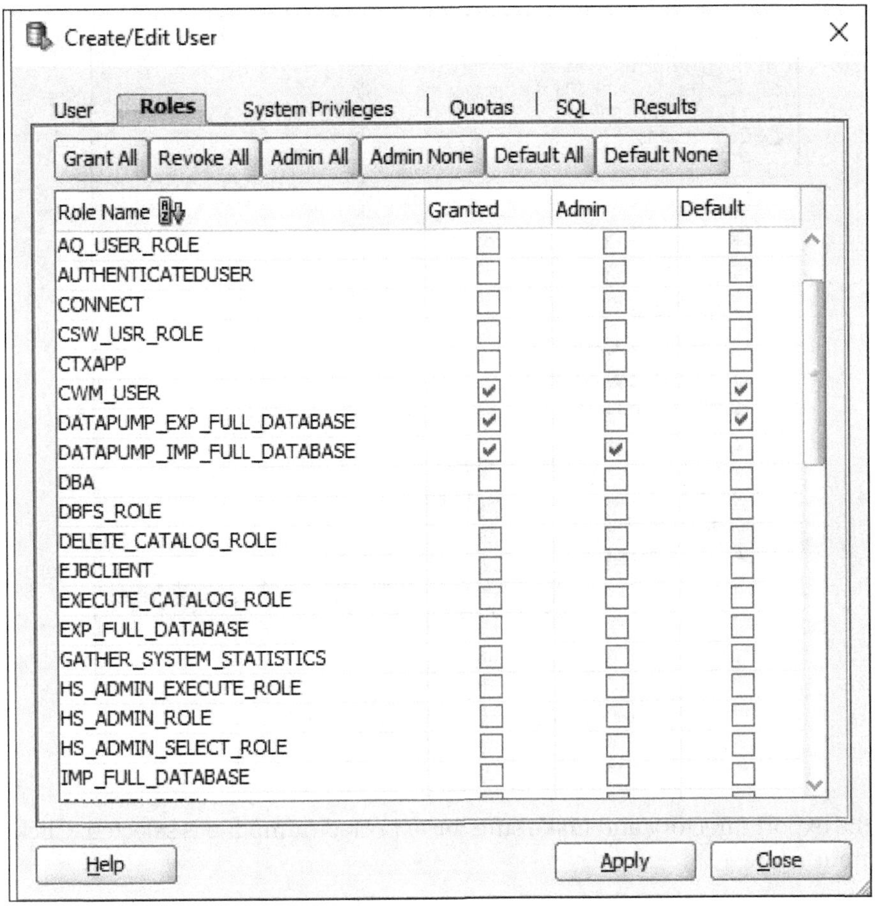

Figure 6-35 Assigning roles to SH_IMP schema

Step 3:
Right-click **Data Pump** -> **Data Pump Import Wizard:**

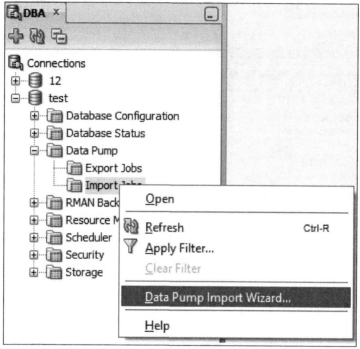

Figure 6-36 Opening Data Pump Import Wizard

Step 4:

Choose the export directory and make sure the exported dump file is selected. Click on the **Next** button:

Figure 6-37 Choosing the exported dump file

Step 5:

Select the **SH** schema in the left pane, then click the ">" button to move the source schema to the right pane. Click on the **Next** button.

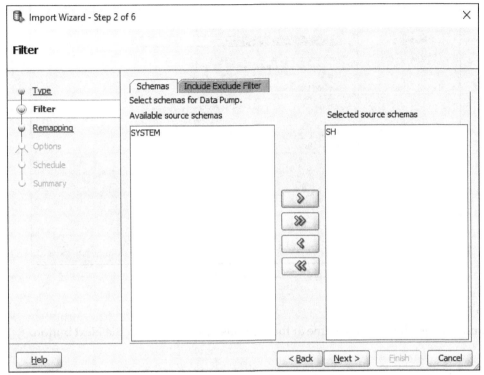

Figure 6-38 Moving the source schema to the right pane

Step 6:

Enter **SH_IMP** at the **Destination** field, then click on the **Next** button:

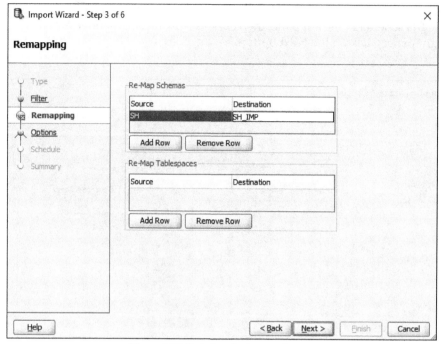

Figure 6-39 Entering the target schema

Step 7:

You can change the LOG file name at the **Options** screen. Click on the **Next** button:

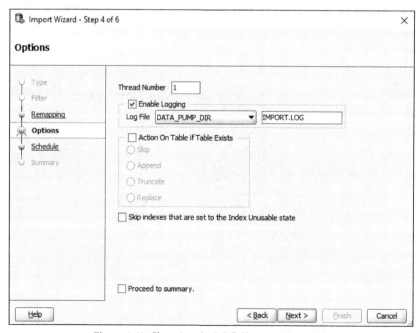

Figure 6-40 Changing the LOG file name if needed

Step 8:

We can set up a time to start tbe job. Here we select to start the job **Immediately**. Click the **Next** button:

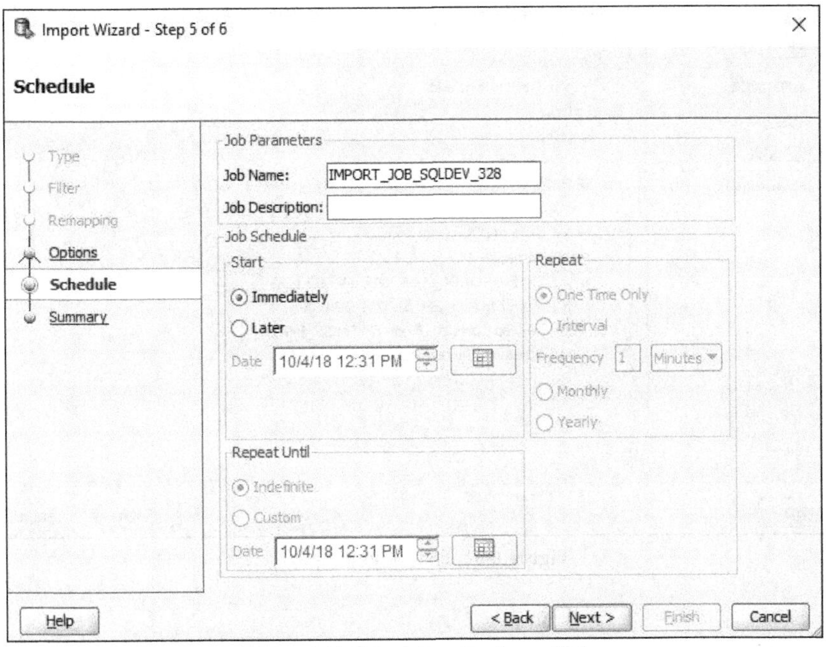

Figure 6-41 Scheduling the import job

Step 9:

You will see the **Summary** information. Click on the **Finish** button:

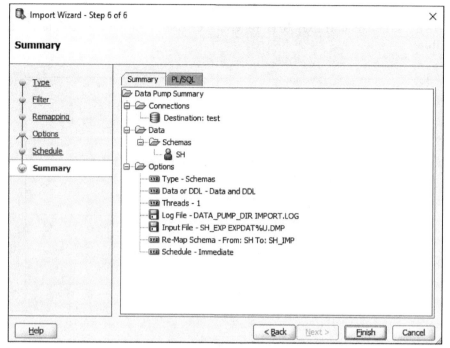

Figure 6-42 The summary page

Step 10:

Check the imported tables and views in the **SH_IMP** schema:

Figure 6-43 The imported tables and views

6.3.2 *Using ESF Database Migration Toolkit*

Prerequisites: Windows XP/Vista/7/8/8.1/10

Oracle 9i and above

Source Server and Database: Oracle 11g SH database

Target Server and Database: Oracle 12c SH database

Migration Tool: ESF Database Migration Toolkit

Steps to migrate from Oracle to Oracle using ESF Database Migration Toolkit:

Step 1:

Install Oracle 12c if you have not done it. Open Oracle SQL Developer and login to Oracle 12c **system** account. Right-click **Other Users** and select **Create User**:

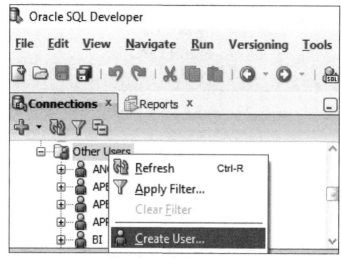

Figure 6-44 Creating a user

Step 2:

Enter a new user name **C##GE** and password. Choose Default Tablespace **USERS** and Temporary Tablespace **TEMP**. Click **Apply** button:

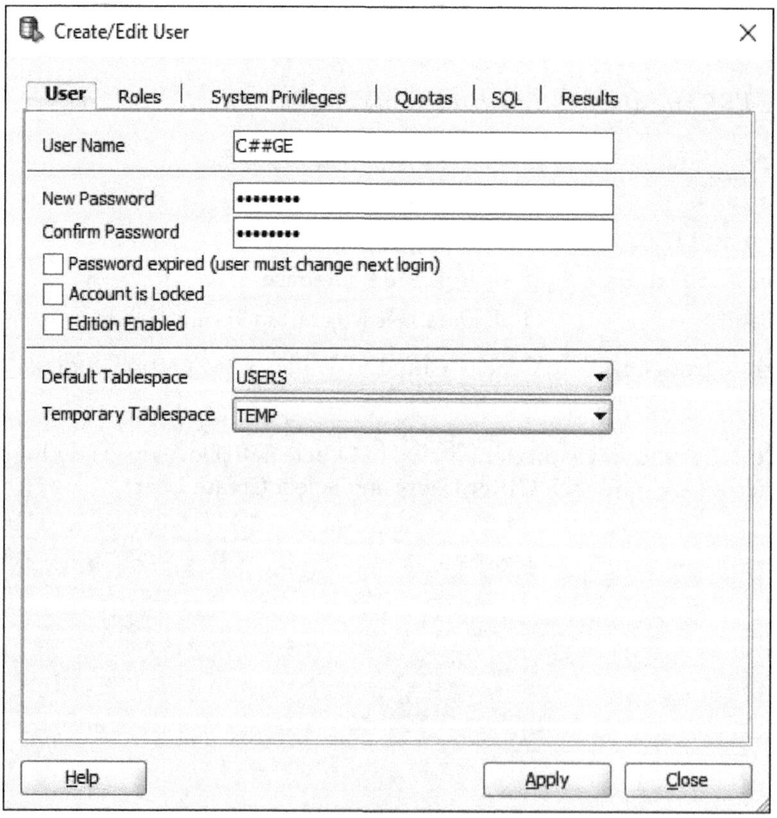

Figure 6-45 New user C##GE

Step 3:

Click **Roles** Tab and check CDB_DBA, CONNECT and DBA roles. Click **Apply** button:

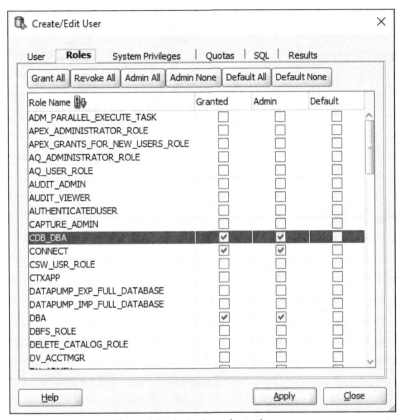

Figure 6-46 Assigning roles to the user

Step 4:

Click **System Privileges** Tab. Check **CREATE SESSION** and **CREATE TABLE**:

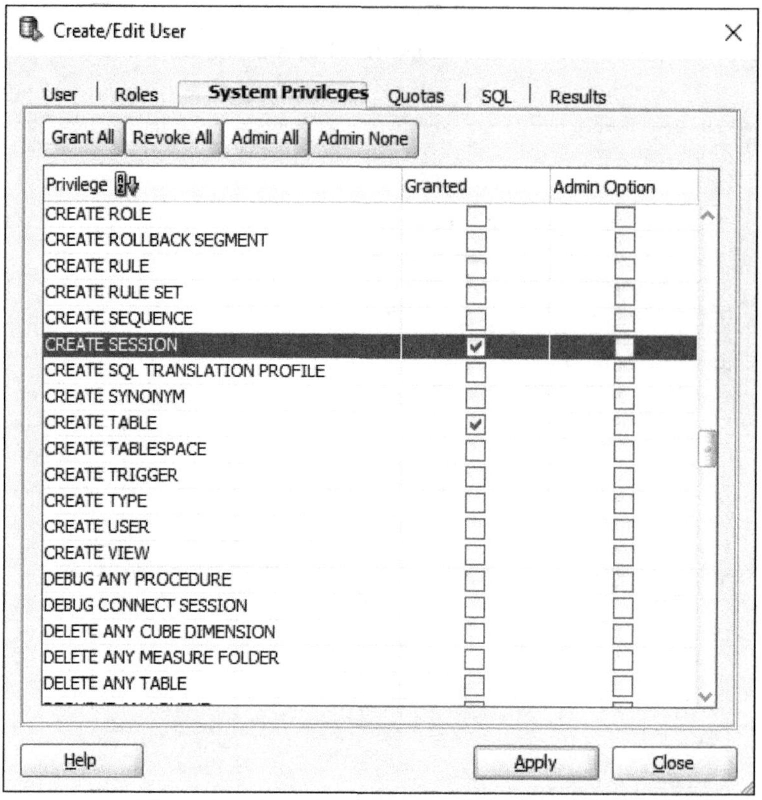

Figure 6-47 Assigning System Privileges to the user

Step 5:

Results Tab shows that user C##GE is created, and permissions are granted.

Figure 6-48 The new user is created

Step 6:

Open ESF Database Migration Toolkit. Choose Oracle 11g **SH** source schema:

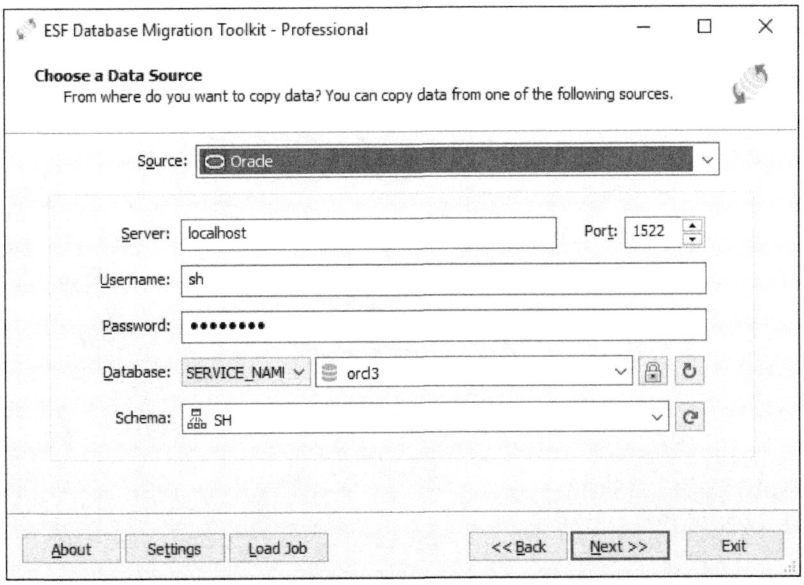

Figure 6-49 Choosing Oracle as source server

Step 7:

Choose Oracle 12c **orcl2** as target database. Enter schema (user) **C##GE**:

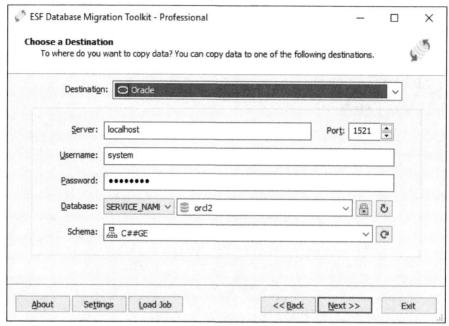

Figure 6-50 Choosing the target database

Step 8:

Click **Next** button. Select tables that need to be migrated.

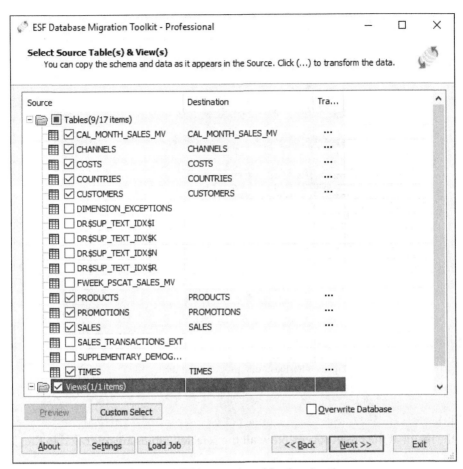

Figure 6-51 Tables are selected for the migration

Step 9:

Click **Next** and then click **Submit** button.

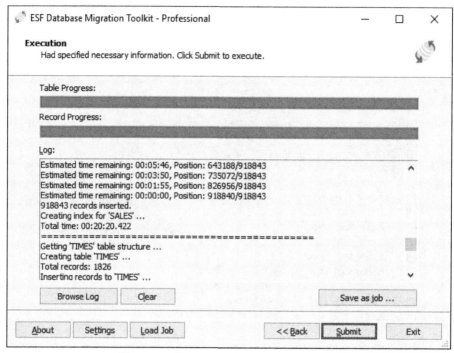

Figure 6-52 Executing progress and the log file

Step 10:

Check the result at the target server. Now all the selected tables are migrated to the C##GE database:

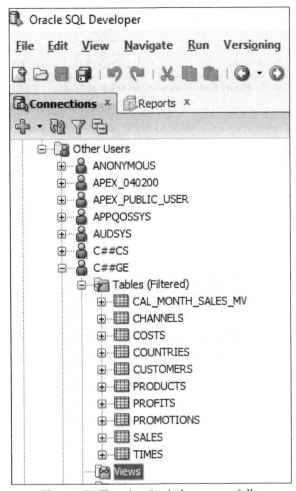

Figure 6-53 The migration is done successfully

Step 11:

Run the following SQL code on Oracle 11g Query window to display Oracle 11g source SH table records: 9198843

```
SELECT
        count (*) as Row_Numbers
FROM
        SH.SALES
```

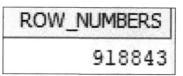

Figure 6-54 The source SH Sales table rows

Step 12:

Run the following SQL code on Oracle 12c Query window to display the target C##GE table sizes:

```
SELECT
        count (*) as Row_Numbers
FROM
        C##GE.SALES
```

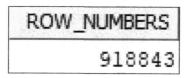

Figure 6-55 The target C##GE Sales table rows

Summary

Chapter 6 covers the following:

- Oracle to MySQL migration example
- Oracle to SQL Server migration example
- Oracle to Oracle migration using Oracle Data Dump Export and Import

Chapter 7

More About Oracle SQL Developer and Oracle Enterprise Manager

We have used Oracle SQL Developer in the last chapter, I would like to introduce other useful features in Oracle SQL Developer and Oracle Enterprise Manager.

7.1 Oracle Query Builder

Oracle SQL Developer has a visual query builder to help developers and DBAs to get query results.

Steps to use the query builder:

Step 1:

Open a database, for example, Oracle HR database and click the **Query Builder** tab.

Step 2:

Drag tables to the query builder screen and select fields that you want to display.

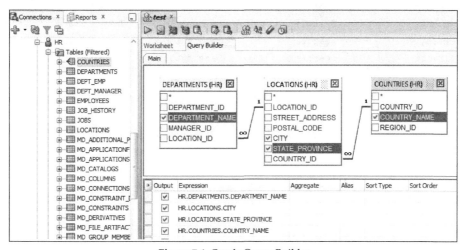

Figure 7-1 Oracle Query Builder

Step 3:

Click on the **Worksheet** tab to see the generated SQL code.

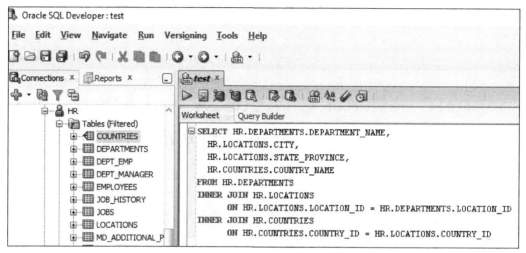

Figure 7-2 SQL code generated by Query Builder

Step 4:

Click the green triangle icon to run the statements:

	DEPARTMENT_NAME	CITY	STATE_PROVINCE	COUNTRY_NAME
1	IT	Southlake	Texas	United States of America
2	Shipping	South San Francisco	California	United States of America
3	Administration	Seattle	Washington	United States of America
4	Purchasing	Seattle	Washington	United States of America
5	Executive	Seattle	Washington	United States of America
6	Finance	Seattle	Washington	United States of America

Figure 7-3 Executing the statement

7.2 Oracle SQL Developer DBA Tool

Oracle SQL Developer allows users with DBA (Database Administrator) privileges to view or do DBA related operations.

- To access the DBA tool, click **View** -> **DBA** on the top menu:

Figure 7-4 Opening DBA tool

- Expend the **Database Status** and click **Status**:

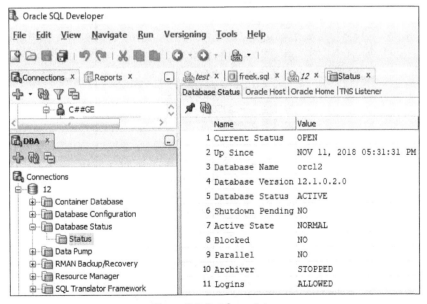

Figure 7-5 Database status

- Click **Security** -> **Roles** to see all the available roles:

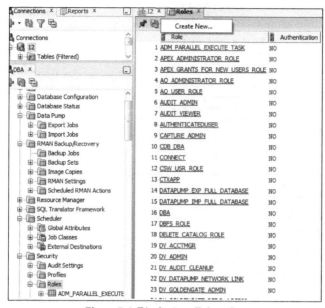

Figure 7-6 Displaying all the roles

- Click **Security** -> **Users** to see all the current users:

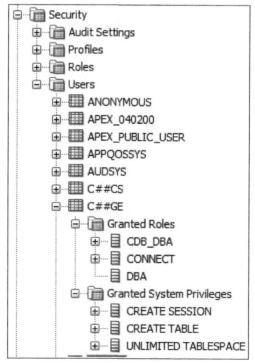

Figure 7-7 Displaying current users

7.3 Oracle 12c Enterprise Manager Database Express

Oracle 12c Enterprise Manager Database Express is a lightweight tool for database administration. It was created when you install the Oracle 12c on your computer. It provides basic tools for storage and performance diagnostics. Comparing with Oracle 11g the 12c the Enterprise Manager Database Express has fewer features. Oracle 12c Enterprise Manager Cloud Control offers many important features for database administrators.

- To access Oracle Enterprise Manager 12c use Google Chrome:

  Enter: https://localhost:5500/em/login

You will see a screen with warning message below. Click on the **Advanced** button then click **Proceed to localhost (unsafe)** link:

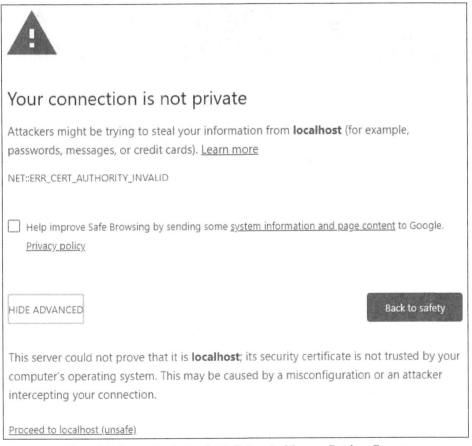

Figure 7-8 Connecting to Oracle 12c Enterprise Manager Database Express

- To login in to Oracle Enterprise Manager 12c, enter **sys** as username and the password. Select the option **"as sysdba":**

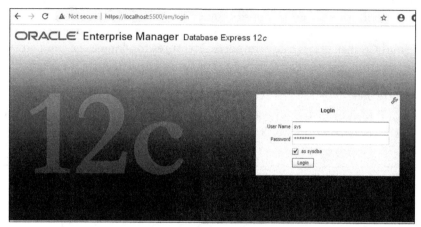

Figure 7-9 Oracle 12c Enterprise Manager Database Express login page

- You can see the database name, the instance name, the Oracle home directory on the left. You can also see performance chart and resources on the right:

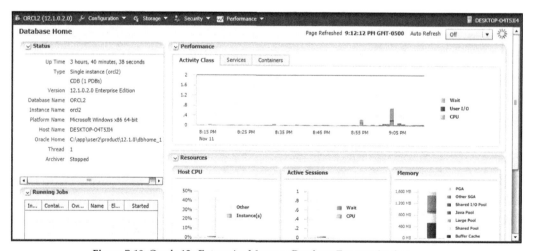

Figure 7-10 Oracle 12c Enterprise Manager Database Express database page

- Click **Configuration** -> **Initialization Parameters to** see the content:

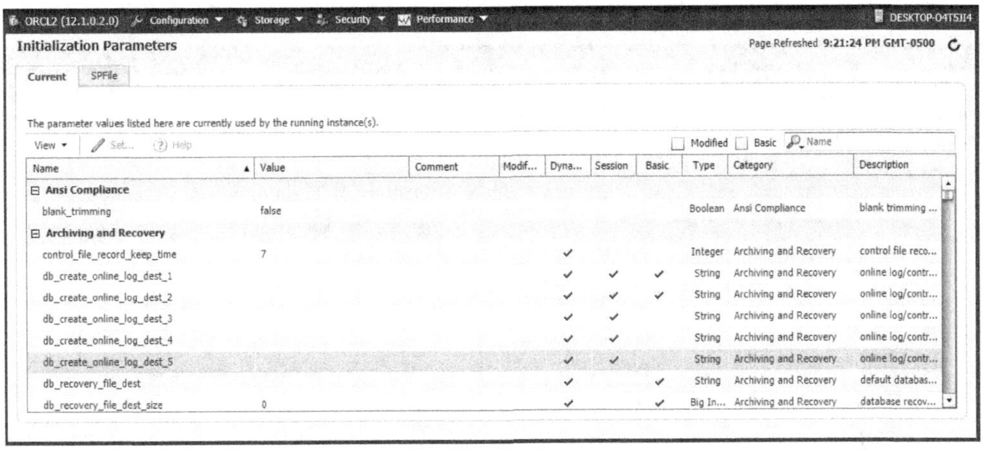

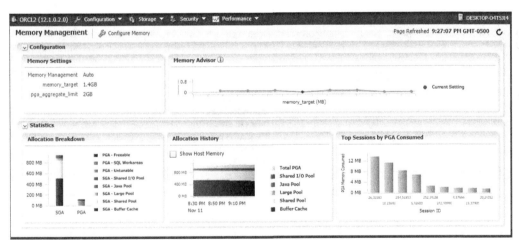

Figure 7-11 Initialization Parameters page

- Click **Configuration** -> **Memory** to see memory setting information and memory advisor chart:

Figure 7-12 Memory Management page

- Click **Storage -> Undo Management** to see undo summary, undo statistics summary and undo advisor chart:

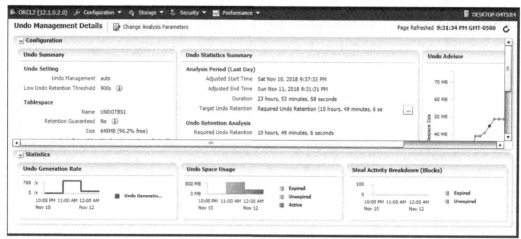

Figure 7-13 Undo Management page

- Click **Storage -> Redo Log Group** to see Redo Log Groups information:

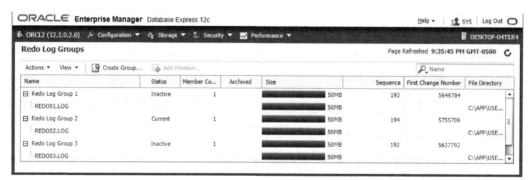

Figure 7-14 Redo Log Groups

- Click **Storage** -> **Control Files** to display control file information:

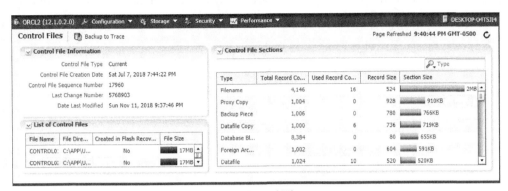

Figure 7-15 Control Files page

- Click **Security** -> **Users** to display current users, account status and expiration dates. The default tablespaces and temporary tablespaces are also displayed. To create a new user, click on the **Create User** tab:

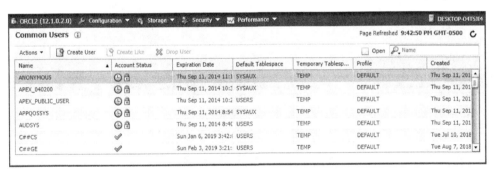

Figure 7-16 Common Users are listed

- Click **Performance** -> **Performance** Hub and select **Summary** tab to see the real time performance charts:

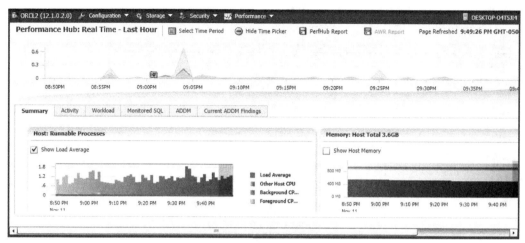

Figure 7-17 Performance Hub page

- Click **Performance** -> **Performance Hub** and select **Activity** tab to see the real time activities charts:

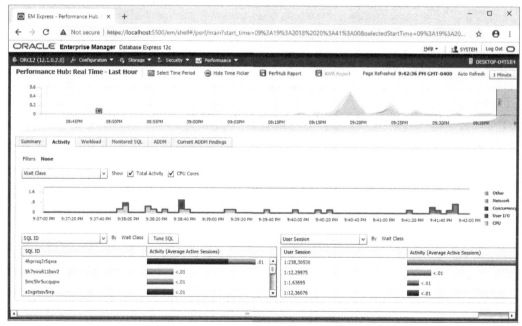

Figure 7-18 Activity tab in Performance Hub page

Summary

Chapter 7 covers the following:

- Using Oracle Query Builder to create a query
- Using Oracle SQL Developer DBA Tool to display roles and users
- Introduction to Oracle 12c Enterprise Manager Database Express

Chapter 8

Microsoft Access Database Migration

Microsoft Access is a powerful desktop application for individual users and small business. However, Microsoft Access can only store up a 2 GB database and it can't handle more than 25 concurrent users. Backup can't be done if Microsoft Access is open.

The main topics in this chapter are illustrated in Figure 8-1: Microsoft Access database migration to MySQL, Microsoft Access database migration to SQL Server and Microsoft Access database migration to Oracle.

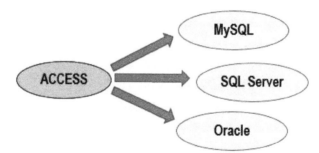

Figure 8-1 Microsoft Access migration paths

8.1 Microsoft Access to MySQL Migration Example

Prerequisites:	Windows XP/Vista/7/8/8.1/10
	Microsoft Access 97 and above
	MySQL 3.23 and above
Source Database:	Microsoft Access 2016 MonthlySalesReports database
Target Server and Database:	MySQL 2016 MonthlySalesReports database
Migration Tool:	ESF Database Migration Toolkit

Steps to migrate Microsoft Access to MySQL:

Step 1:

Open ESF Database Migration Toolkit. Choose Microsoft Access (*.mdb) as source server. Enter database name: MonthlySalesReports.mdb

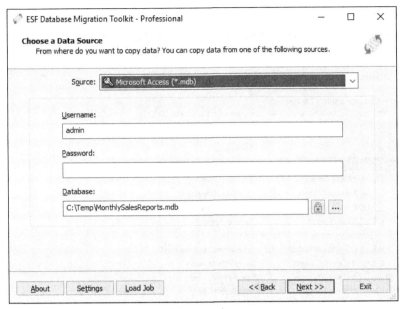

FIGURE 8-2 CHOOSING MICROSOFT ACCESS AS SOURCE DATABASE

Step 2:

Choose **MySQL** as a target server. Enter the default server name **localhost** and default port **3306**. Enter username **root** and password. Enter database name **MonthlySalesReports.**

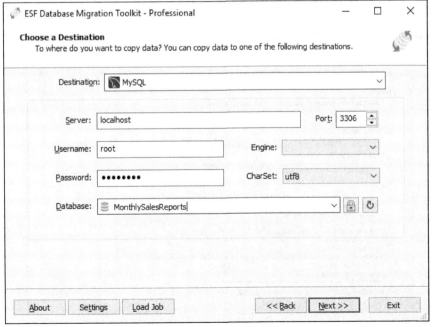

FIGURE 8-3 CONNECTING TO MYSQL SERVER

Step 3:

Click **Next**. Select source tables and views that need to be migrated:

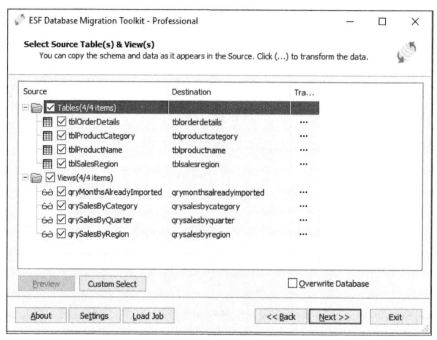

Figure 8-4 Selecting tables and views for migration

Step 4:

Click **Next** then click on **Submit** button.

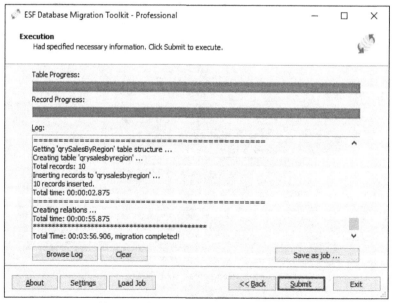

Figure 8-5 Migration progress and the log

Step 5:

After the migration we can check the source tblProductName table records: 266 rows.

Figure 8-6 Displaying tblProductName table records in Microsoft Access

Step 6:

Open MySQL Workbench and check target tblProductName table records: 266 rows.

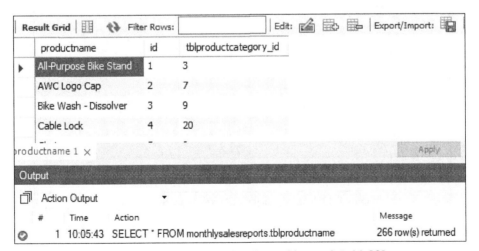

Figure 8-7 Displaying tblProductName table records in MySQL

8.2 Microsoft Access to SQL Server Example

Prerequisites:	Windows XP/Vista/7/8/8.1/10
	Microsoft Access 97 and above
Source Database:	Microsoft Access MonthlySalesReports database
Target Server and Database:	SQL Server 2016 MonthlySalesReports database
Migration Tool:	ESF Database Migration Toolkit

Steps to migrate Microsoft Access to SQL Server:

Step 1:

Open ESF Database Migration Toolkit.

- Choose Microsoft Access (*.mdb) as source database
- Enter database name: MonthlySalesReports.mdb

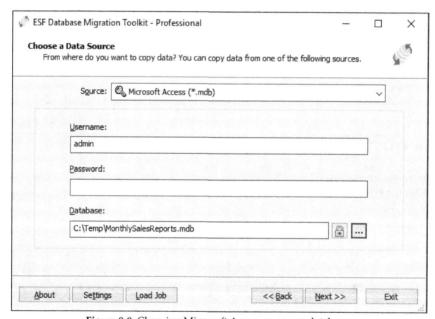

Figure 8-8 Choosing Microsoft Access as source database

Step 2:

Choose **SQL Server (Windows Authentication)** as a target server. Enter the server name and database name **MonthlySales.**

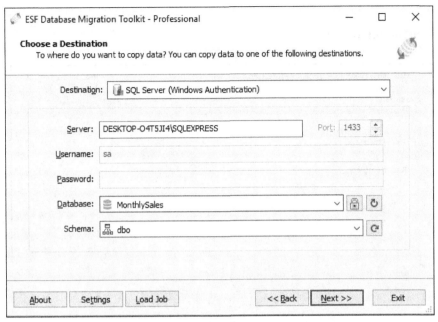

Figure 8-9 Choosing SQL Server as target server

Step 3:

Select all the source tables and Click on the **Next** button:

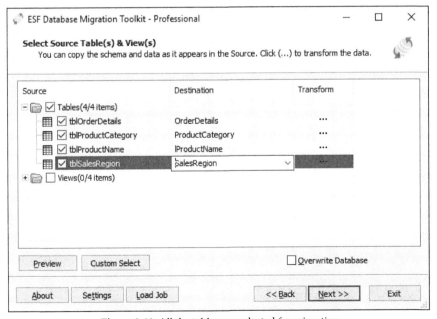

Figure 8-10 All the tables are selected for migration

Step 4:

Click on the **Submit** button.

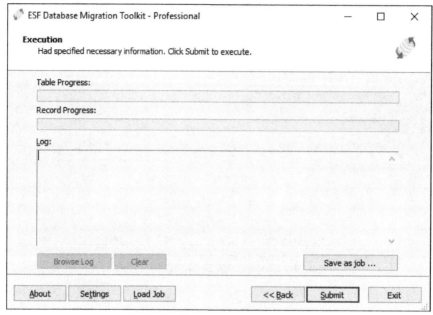

Figure 8-11 Executing the migrating task

Step 5:

Verify the migration has no error message:

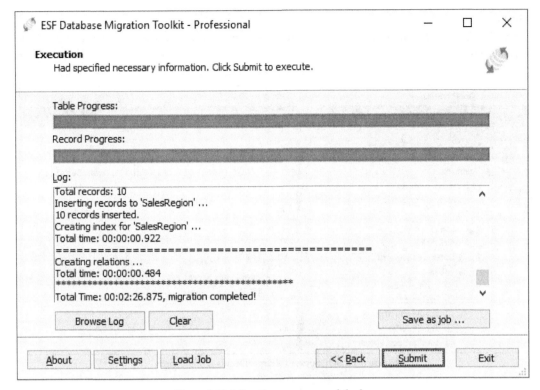

Figure 8-12 Migration progress and the log

Step 6:

The source tblOrderDetails table has 121,317 rows.

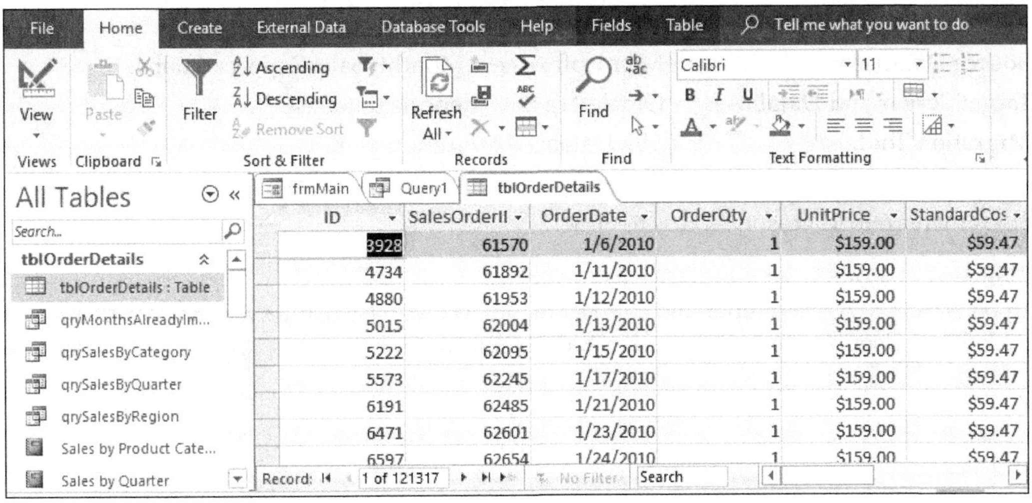

Figure 8-13 Microsoft Access tblOrderDetails table records

Step 7:

Open SQL Server Management Studio you can verify that target tblOrderDetails table has the same records.

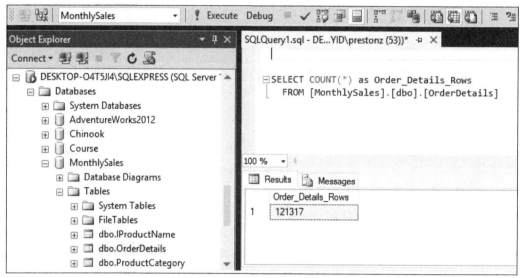

Figure 8-14 SQL Server tblOrderDetails table records

8.3 Microsoft Access to Oracle Migration Example

Prerequisites:	Windows XP/Vista/7/8/8.1/10
	Microsoft Access 97 and above
	Oracle 9i and above
Source Database:	Microsoft Access MonthlySalesReports database
Target Server and Database:	Oracle 11g SalesReports database
Migration Tool:	ESF Database Migration Tookit

Steps to migrate Microsoft Access to Oracle:

Step 1:

Open **Oracle SQL Developer** and connect to Oracle database using **system** account. Right-click Other Users to create a new user: **SalesReports**.

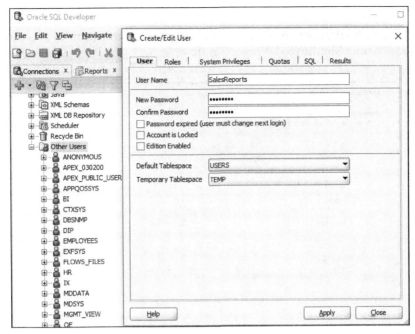

Figure 8-15 Creating a new user

Step 2:

Grant the **SalesReports** user system privileges:

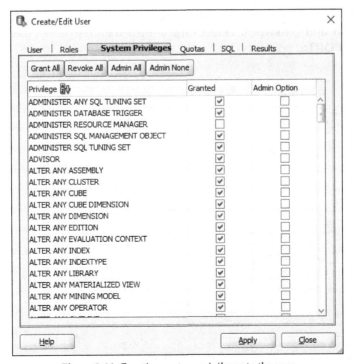

Figure 8-16 Granting system privileges to the user

Step 3:

Open ESF Database Migration Toolkit. Choose Microsoft Access (*.mdb) as source server. Enter database name: MonthlySalesReports.mdb

Figure 8-17 Choosing Microsoft Access as source database

Step 4:

Choose **Oracle** as a target server. Enter the default server name **localhost.** Enter username **system** and password. Select Oracle orc3 database and enter schema name **MONTHLY_REPORTS**.

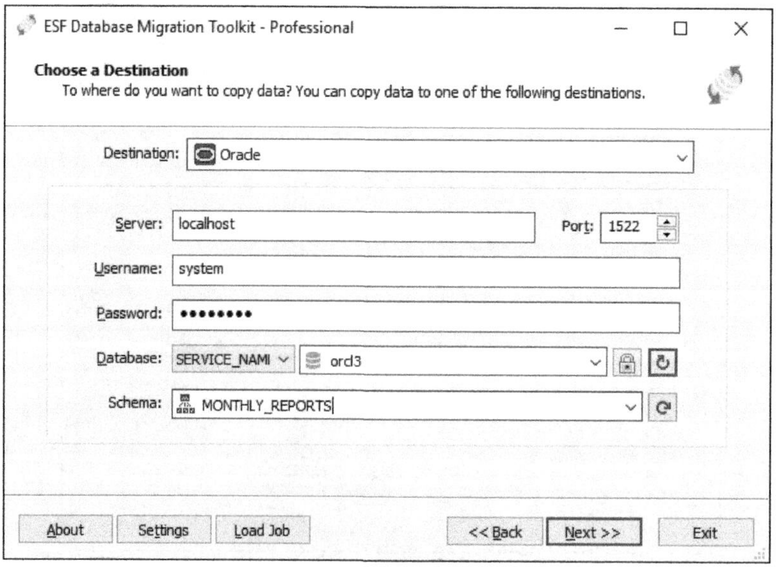

Figure 8-18 Providing target server information

Step 5:

Select all the source tables:

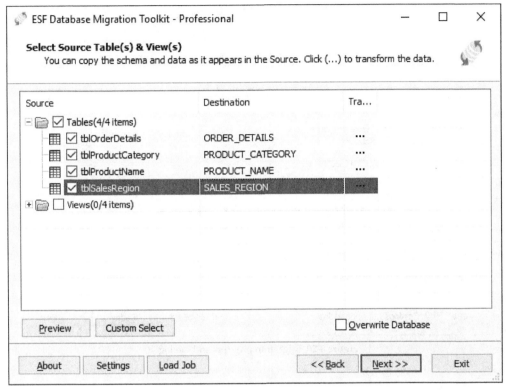

Figure 8-19 All the source tables are selected

Step 6:

Click on **Next** button:

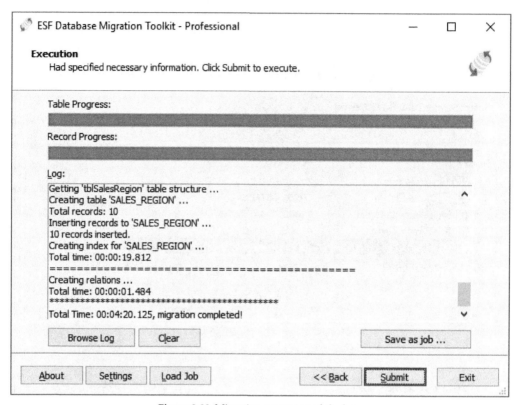

Figure 8-20 Migration progress and the log

Step 7:

After the migration let us check the source tbllOrderDetails table records: 121,317.

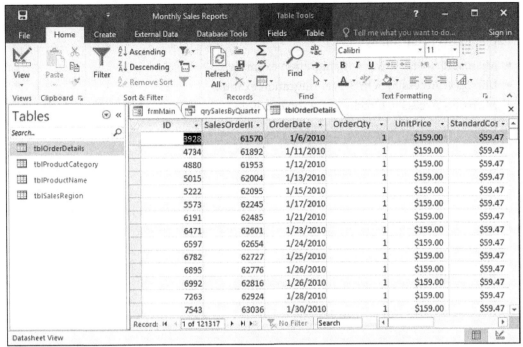

Figure 8-21 Microsoft Access tblOrderDetails table records

Step 8:

Figure 8-22 shows the query result that matches the Microsoft Access table rows:

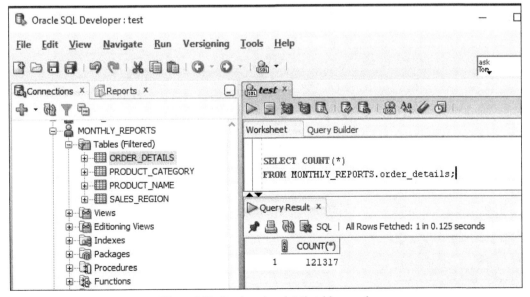

Figure 8-22 Oracle order_details table records

Summary

Chapter 8 covers the following:

- Microsoft Access to MySQL migration example
- Microsoft Access to SQL Server migration example
- Microsoft Access to Oracle migration example

Index

About the Author

Preston Zhang has over 20 years of experiences in database design and implementation. As a database administrator, he manages Oracle, SQL Server and MySQL database servers for university departments in Georgia. He has done many database migration projects in Oracle, SQL Server and MySQL database systems. He has written many queries in Oracle SQL, SQL Server T-SQL and MySQL to process millions of records for business reports. He has developed Web applications using Oracle database as back-end for a large health care company. He has taught undergraduate database and programming courses in private universities for over 10 years. He has a Master of Science degree in Computer Information Systems from University of Wisconsin-Parkside. He lives in Georgia with his family and can be reached at prestonz668@gmail.com.

9781138391628